ADVANCES IN EVOLUTION EQUATIONS

EVOLUTIONARY PROCESSES AND APPLICATIONS

ADVANCES IN EVOLUTION EQUATIONS

GASTON M. N'GUÉRÉKATA - SERIES EDITOR –

MORGAN STATE UNIVERSITY, BALTIMORE, MD, US

Evolutionary Processes and Applications
Gaston M. N'Guérékata (Editor)
2019. ISBN: 978-1-53616-350-6

Evolution Equations: Almost Periodicity and Beyond (In Memory of Professor V. V. Zhikov)
Gaston M. N'Guérékata, Jin Liang and Alexander Pankov (Editors)
2018. ISBN: 978-1-53614-258-7

ADVANCES IN EVOLUTION EQUATIONS

EVOLUTIONARY PROCESSES AND APPLICATIONS

GASTON MANDATA N'GUÉRÉKATA
EDITOR

nova
science publishers
New York

Library of Congress Cataloging-in-Publication Data

Names: N'Guerekata, Gaston M., 1953- editor.
Title: Evolutionary processes and applications / [edited by Gaston M. N'guérékata].
Description: New York : Nova Science Publishers, Inc., [2019] | Series: Advances in evolution equations | Includes bibliographical references and index. |
Identifiers: LCCN 2019041475 (print) | LCCN 2019041476 (ebook) | ISBN 9781536163506 (hardcover) | ISBN 9781536163513 (adobe pdf)
Subjects: LCSH: Evolution equations.
Classification: LCC QA377.3 .E96 2019 (print) | LCC QA377.3 (ebook) | DDC 515/.39--dc23
LC record available at https://lccn.loc.gov/2019041475
LC ebook record available at https://lccn.loc.gov/2019041476

Published by Nova Science Publishers, Inc. † New York

CONTENTS

Preface		**vii**
Chapter 1	Traveling Waves in Discrete Models of Biological Populations with Sessile Stages *Ahlam H. Alzahrani*	**1**
Chapter 2	Generation and Evolution of Magneto-Acoustic Waves in an Isothermal Atmosphere (I) *H. Y. Alkahby, D. A. Rossmanith, R. A. Shulaiba and D. R. Jagessar*	**15**
Chapter 3	Ergodic-Type Theorem for Quantum Dynamical Semigroup *Michael O. Ogundiran*	**37**
Chapter 4	On the Stability of Higher Order Differential Equations *B. S. Kalitine*	**49**
Chapter 5	Asymptotics Related to Nonlinear Effects for a Class of Higher Order Parabolic Equations *Mahmoud Qafsaoui*	**61**
Chapter 6	Discriminating Sentinel with Given Sensitivity for the Heat Equation with Fourier Boundary Condition *Sadou Tao and Elisée Gouba*	**77**

Chapter 7 Local Stable Manifolds of Admissible Classes for Parabolic Functional Equations and Applications to Hutchinson Models **105**
Nguyen Thieu Huy and Dinh Xuan Khanh

Chapter 8 Stackelberg Control of a Backward Linear Heat Equation **127**
L. L. Djomegne Njoukoue, G. Mophou and G. Deugoue

Chapter 9 *S*-Asymptotically ω-Periodic Mild Solutions for Second Order Evolution Equations **151**
Mouffak Benchohra, Gaston M. N'Guérékata and Noreddine Rezoug

Chapter 10 The Initial Boundary Value Problem of an Evolutionary $p(x)$-Laplacian Equation **175**
Zhao Jiangyan and Huashui Zhan

Chapter 11 On a Criterion for Regularity of Leray Weak Solutions to the Navier - Stokes Equations Involving One Directional Derivative **205**
Ngo Van Giang and Nguyen Minh Tri

About the Editor **215**

Index **219**

Related Nova Publications **223**

PREFACE

This book presents and discusses new developments in the study of evolutionary processes. Topics discussed include evolution of magneto-acoustic waves in isothermal atmosphere, quantum dynamical semigroups, traveling waves in discrete models of biological population, motion of electrorheological fluids, Stacklberg control of a backward linear heat equation, Leray weak solutions of Navier-Stokes equation involving one directional derivative, and initial value boundary problem of an evolutionary p(x)-Laplacian equation.

Chapter 1 deals with the existence of traveling waves for a discrete model of biological population with sessile stages. The authors discuss the existence of traveling waves when the migration probability may not have density, or if it has density function, the density function may not be continuous. For this end, the authors impose some conditions that guarantee the monotone property of the evolution operator, so that they use the Monotone Iteration Method.

In Chapter 2, the authors will study generation, propagation, reflection and transmission of magneto-acoustic waves in an isothermal solar atmosphere under the effect of arbitrary values of the effect of thermal conduction, viscosity and horizontal magnetic field. A general solution was obtained. It is shown that if the effect thermal conduction dominates the wave generation and propagation, the solar atmosphere can be divided into two different regions which are connected by an absorbing and reflecting layer. Below the reflecting layer, the oscillatory process is adiabatic and the solution can be written as a linear combination of upward and down ward propagating waves. Above the reflecting layer, the solution decays exponentially with the altitude. Also, if the combined effects of viscosity and magnetic field dominate the wave's generation, the atmosphere can be divided into two different regions which are connected by an absorbing and reflecting layer and the effect of thermal conduction can be eliminated. In both cases the reflection coefficient was determined. The general solution was compared to another solution in which the inner and the outer approximate solutions were matched in an overlapping domain. The results of the two solutions were investigated in connection with the heating mechanisms of the solar atmosphere.

In Chapter 3, the authors shall be concerned with quantum dynamical semigroup on von Neumann algebra. The existence result for a long-time asymptotic behaviour of quantum dynamical semigroup of certain open quantum systems was established. The authors first established the existence of solution for Cauchy problem involving an infinitesimal generator of such semigroup then an ergodic-type theorem for the quantum dynamical semigroup was established.

In Chapter 4, the authors consider the problem of stability of the equilibrium of a non-linear differential equation of order n. The investigations are carried out using the method of semi-definite Lyapunov's functions for the problems of stability, asymptotic stability (local and global), and instability. The method of constructing the semi-definite functions, suitable for solving problems of stability, is implicitly stated. The results of asymptotic stability of the equilibrium coincide with necessary and sufficient conditions in the corresponding linear case. Consequently, they meet generally accepted requirements. The conducted studies show that the use of semi-defined positive functions can give advantages in comparison with the classical method of application of Lyapunov's definite positive functions.

Chapter 5 deals with the large time behavior of solutions to the initial value problem related to the class of nonlinear parabolic equations of type

$$\begin{cases} u_t + \Delta^m(a\Delta^m u) = \nabla^\theta.\mathcal{J}(\nabla^\mu u), \\ u(.,0) = u_0 \in L^p(\mathbb{R}^n). \end{cases} \tag{1}$$

The vector field $\mathcal{J}$ represents a nonlinearity and the bounded measurable non-negative function a is independent of time.

Chapter 6 studies the null controlability problem with two constraints on a pair of controls. The authors apply these results to a discriminating sentinel with given sensitivity to detect some parameters in a pollution problem governed by a semi-linear parabolic equation with Fourier boundary condition.

Chapter 7 proves the existence of local stable manifolds of admissible classes for solutions to non-autonomous parabolic functional equation of the form $\dot{u}(t) = A(t)u(t) + f(t, u_t)$, $t \geq 0$. The special feature of this kind of manifolds is that they are constituted by trajectories of the solutions belonging to admissible function spaces. The existence of these manifolds obtained in the case that the family of operators $(A(t))_{t\geq 0}$ generate the evolution family $(U(t,s))_{t\geq s\geq 0}$ having an exponential dichotomy on the half-line and the nonlinear forcing term f satisfies the local φ-Lipschitz condition, i.e., $\|f(t,\phi) - f(t,\psi)\| \leq \varphi(t)\|\phi - \psi\|_C$ where ϕ, ψ belong to some ball in $C = C([-r, 0], X)$, $\varphi(\cdot)$ is in some admissible Banach function space and satisfies some certain conditions. The authors then apply the results to Hutchinson model to show the local admissibly stable manifold for Hutchinson equations.

The authors of Chapter 8 are interested in the hierarchic control problem for a linear backward heat equation. They assume that they can act on the system via two distributed controls. One control called follower solves an optimal control problem which consists in bringing the flux of the system to zero. The other control called leader solves a null controllability problem. The results are achieved by means of inequality of observability associated to a nonhomogeneous Dirichlet boundary condition.

Chapter 9 is devoted to study the existence of S-asymptotically ω-periodic mild solution to some classes of second order semilinear evolution equation in Banach space, via the techniques of measure of noncompactness. The investigation is based on a new fixed point result which is a generalization of the well known Darbo's fixed point theorem. Finally, an example is presented to illustrate the main findings.

In Chapter 10 is considered the initial boundary value problem of an equation

$$u_t = div(|\nabla u|^{p(x)-2}\nabla u) + f(x,t,u).$$

The problems describe the motion of electrorheological fluids. If the variable exponent $p(x)$ is required to satisfy a logarithmic Hölder continuity condition, there are many references to study the equation. If the exponent $p(x)$ does not satisfy the logarithmic Hölder continuity condition. the related references are much fewer. If $\inf p(x) > 2$, Lian et al. proved the existence and uniqueness of the equation. In the chapter, by difference method, the case of the equation with $2 > ess\,\sup p(x) \geq ess\,\inf p(x) > 1$ is studied. The existence and uniqueness of weak solutions are obtained provided that $ess\,\inf p(x) > \frac{2N}{N+1}$.

In Chapter 11, the authors obtain sufficient conditions for the regularity of

Leray solutions of Navier-Stokes equations (NSE) in three dimensional space. The authors prove that if the third derivative of the velocity $\frac{\partial u}{\partial x_3}$ belongs to the space $L_t^{s_0} L_x^{r_0}$, where $\frac{2}{s_0} + \frac{3}{r_0} \leq \frac{5}{3} + \frac{10}{3(9r_0-2)}, \quad r_0 \geq 3$ then the solution is regular.

In: Evolutionary Processes and Applications
Editor: G. M. N'Guérékata
ISBN: 978-1-53616-350-6

Chapter 1

TRAVELING WAVES IN DISCRETE MODELS OF BIOLOGICAL POPULATIONS WITH SESSILE STAGES

Ahlam H. Alzahrani*
Department of Mathematics and Statistics
University of Arkansas at Little Rock, AR, US

Abstract

This chapter deals with the existence of traveling waves for a discrete model of biological population with sessile stages. We discuss the existence of traveling waves when the migration probability may not have density, or if it has density function, the density function may not be continuous. For this end, we impose some conditions that guarantee the monotone property of the evolution operator, so that we use the Monotone Iteration Method.

1. INTRODUCTION AND PRELIMINARY

Mathematical models of biological invasions can be written as reaction-diffusion equations as in [1], and [2] or, more recently, as integrodifference and integrodifferential equations as in [3]. Studying about the existence of a

*Corresponding Author's Email: ahalzahrani@ualr.edu.

spreading speed and traveling waves uses a recursion of the form

$$u_{n+1} = Q[u_n], \quad n \in \mathbb{N}$$

where u_n represents the density of the population in the habitat (which we assume to be $\mathbb{R}^d$) at time $n \in \mathbb{N}$. There are many papers, including [1-12], on the existence of traveling waves and spreading speed in population models that provide important insights into population behavior. In addition, Lutcsher and Nguyen in [4] studied the existence of traveling waves in discrete model such that the probability for migration satisfies a compactness condition if the kernel is continuous. Le and Nguyen in [5] studied the existence of traveling waves in a general discrete model for populations such that the dispersal probability has no continuous functions. On the other hand, Li in [6] shows that there may not be traveling wave solutions in plant population when the recursion operator in the model is not compact. In this chapter, we will study the existence of traveling waves in discrete models of biological populations with sessile stages, and we do not require the probability for migration to have density function, and even if it has we do not require its density function to be continuous.

The equations we consider in this work are formulated in discrete time since natural annual variation often imposes clearly distinct seasons of growth and spread of individuals. As a general model for second generation of sessile stages we consider in this chapter is the following:

$$\begin{cases} u_{n+1} = \sum_{n=1}^{\infty} \eta_i f(u_n(x - c_i)) + \alpha \int_{-\infty}^{\infty} K(x - p_i - y) f(u_n(y))dy + \beta w_n(x - p_2), \\ w_{n+1} = \sum_{n=1}^{\infty} \rho_i f(w_n(x - c_i)) + \gamma \int_{-\infty}^{\infty} K(x - q_i - y) f(w_n(y))dy + \delta u_n(x - q_2), \end{cases} \tag{1.1}$$

where the constants satisfy $\alpha, \beta, \gamma, \delta, p_i, q_i \geq 0$ and

$$\begin{aligned} \alpha + \gamma + \sum_{i=1}^{\infty} \eta_i &< 1, \\ \beta + \delta &< 1, \end{aligned}$$

u_n represents the density of seedlings at time n, w_n represents the density of seeds at time n, and the kernal K is the density of probability measure on $\mathbb{R}$, c_i is a constant for each i=1,2..., and f is a scalar function that characerizes the growth of the population. An example for such a function is the Beverton-Holt function:

$$f(u) = \frac{ru}{1+(r-1)u}, \tag{1.2}$$

where $r > 1$.

We will start by assuming the equilibrium points and finding the wave equations. Then, we give details in studying spreading speeds by the general theory of spreading speed that developed by Weinberger in [12]. At the end, we study the existence of traveling waves by the Monotone Iteration Method.

1.1. Notations and Assumptions

We will see $BM(\mathbb{R}, \mathbb{R}^2)$ ($BC(\mathbb{R}, \mathbb{R}^2)$), which means the space of all measurable and bounded real valued functions on $\mathbb{R}$ (the space of all bounded continuous real valued functions on $\mathbb{R}^2$, respectively) with sup-norm. For given positive M_1, M_2, we dine $C_M := CM(\mathbb{R}, [0, M]) := \{f \in CM(\mathbb{R}, \mathbb{R}^2) | f(x) \in [0, M_1] \times [0, M_2]\}$, and $B_M := BM(\mathbb{R}, [0, M]) := \{f \in BM(\mathbb{R}, \mathbb{R}^2) | f(x) \in [0, M_1] \times [0, M_2]\}$. The metric on C_M is defined by the sup norm. In $BM(\mathbb{R}, \mathbb{R}^2)$ we use the natural order defined as $u \leq v$ if and only if $u_i(x) \leq v_i(x)$ for all $x \in \mathbb{R}, i = 1, 2$, where $u(x) = (u_1(x), u_2(x))^T, v(x) = (v_1(x), v_2(x))^T$.

2. MAIN RESULTS

In this chapter we will assume that the density $K(\cdot)$ satisfies the following properties:

Standing Assumption on K

1. $K(\cdot)$ is non-negative, and measurable on $\mathbb{R}$,
2. For all $\mu \in \mathbb{R}$

$$\int_{-\infty}^{\infty} K(x)dx = 1, \quad \int_{-\infty}^{\infty} K(x)e^{\mu x}dx < \infty,$$

3. For all $x_0 \in \mathbb{R}$, the following is true

$$\lim_{x \to x_0} \int_{-\infty}^{\infty} |K(x-y) - K(x_0 - y)|dy = 0.$$

2.1. Wave Equations

We assume throughout this section that Eq.(1.1) has two equilibria $(0,0)^T$ and $(M_1, M_2)^T$ with $M_1, M_2 > 0$, that is,

$$\begin{cases} M_1 = \sum_{n=1}^{\infty} \eta_i f(M_1) + \alpha \int_{-\infty}^{\infty} K(x - p_1 - y) f(M_1) dy + \beta M_2, \\ M_2 = \sum_{n=1}^{\infty} \rho_i f(M_2) + \gamma \int_{-\infty}^{\infty} K(x - q_1 - y) f(M_2) dy + \delta M_1. \end{cases}$$

$$\begin{cases} M_1 = f(M_1) \sum_{n=1}^{\infty} \eta_i + \alpha f(M_1) \int_{-\infty}^{\infty} K(x - p_1 - y) dy + \beta M_2, \\ M_2 = f(M_2) \sum_{n=1}^{\infty} \rho_i + \gamma f(M_2) \int_{-\infty}^{\infty} K(x - q_1 - y) dy + \delta M_1. \end{cases}$$

Since

$$\int_{-\infty}^{\infty} K(x - p_1 - y) dy = 1,$$

we have

$$\begin{cases} M_1 = (\alpha + \sum_{n=1}^{\infty} \eta_i) f(M_1) + \beta M_2, \\ M_2 = (\gamma + \sum_{n=1}^{\infty} \rho_i) f(M_2) + \delta M_1. \end{cases}$$

Example 1. *In equation (1.1), we choose the Beverton-Holt function defined by (1.2) for f. Then, it is easy to see that $(M_1, M_2)^T$ is a positive equilibrium of Eq.(2.2), if we choose*

$$M_1 = (\frac{BrM_2}{(1 + (r-1)M_2} + \delta M_2)\frac{BC}{C - Ar}, \qquad M_2 = \frac{M_1(C - Ar)}{BC}$$

such that:

$$A = \alpha + \sum_{n=1}^{\infty} \eta_i, \qquad B = \gamma + \sum_{n=1}^{\infty} \eta_i, \qquad C = 1 + (r-1)M_1.$$

Definition 2. *Let $\phi(\xi) := (\varphi(\xi), \psi(\xi))^T$ be an ordered pair of continuous scalar functions on $\mathbb{R}$. It is called a* wave solution *of Eq.(1.1) if*

$$(u_n(x), w_n(x))^T := (\varphi(x - nc), \psi(x - nc))^T$$

is a continuous solution of (1.1) such that

$$\lim_{x \to +\infty} \phi(x) = (0,0)^T, \qquad \lim_{x \to -\infty} \phi(x) = (M_1, M_2)^T.$$

If we substitute

$$(u_n(x), w_n(x))^T := (\varphi(x - nc), \psi(x - nc))^T$$

into Eq.(1.1) and set $\xi := x - nc - c$, we will come up with

$$\begin{cases} \varphi(\xi) = \sum_{n=1}^{\infty} \eta_i f(\varphi(\xi - nc - c_i)) + \alpha \int_{-\infty}^{\infty} K(\xi + nc + c - p_1 - y) f(\varphi(y - nc)dy + \beta\psi(x - nc - p_2), \\ \psi(\xi) = \sum_{n=1}^{\infty} \rho_i f(\psi(\xi - nc - c_i)) + \gamma \int_{-\infty}^{\infty} K(\xi + nc + c - q_1 - y) f(\psi(y - nc)dy + \delta\psi(x - nc - q_2) \end{cases} \tag{2.1}$$

Equation (2.1) will be referred to as *the wave equation* associated with Eq. (1.1).

In order to study the wave equation (2.2), we introduce an operator S acting on $BM(\mathbb{R}, [0, M_1] \times [0, M_2])$ defined as

$$S : \begin{bmatrix} \varphi(\xi) \\ \psi(\xi) \end{bmatrix} \mapsto \begin{bmatrix} \sum_{n=1}^{\infty} \eta_i f(\varphi(\xi - nc - c_i)) + \alpha \int_{-\infty}^{\infty} K(\xi + nc + c - p_1 - y) f(\varphi(y - nc)dy + \beta\psi(x - nc - p_2) \\ \sum_{n=1}^{\infty} \rho_i f(\psi(\xi - nc - c_i)) + \gamma \int_{-\infty}^{\infty} K(\xi + nc + c - q_1 - y) f(\psi(y - nc)dy + \delta\psi(x - nc - q_2) \end{bmatrix}$$

With this notation, we have defined a dynamical system in $BM(\mathbb{R}, [0, M_1] \times [0, M_2])$ that leaves $BC(\mathbb{R}, [0, M_1] \times [0, M_2])$ invariant, namely

$$\phi_{n+1} = S[\phi_n], \quad \phi \in BM(\mathbb{R}, [0, M]), \ \ n \in \mathbb{N}. \tag{2.2}$$

For a given non-negative constant c, we define

$$S_c[\phi](x) := ST_{-c}(x) = S[\phi(\cdot + c)](x), \qquad \phi \in BM(\mathbb{R}, [0, M_1] \times [0, M_2]).$$

Definition 3. *The dynamical system (2.1) is said to be* monotone *if it is order-preserving with respect to the natural order in $BM(\mathbb{R}, [0, M])$, that is,*

$$S[\phi] \leq S[\psi],$$

whenever $\phi, \psi \in BM(\mathbb{R}, [0, M_1] \times [0, M_2])$ such that $\phi \leq \psi$.

For a fixed positive constant c, let us consider the operators

$$Q_c : \begin{bmatrix} \varphi \\ \psi \end{bmatrix} (\xi) \mapsto \begin{bmatrix} \alpha \int_{-\infty}^{\infty} K(\xi + nc + c - p_1 - y) f(\varphi(y - nc)dy \\ \gamma \int_{-\infty}^{\infty} K(\xi + nc + c - q_1 - y) f(\psi(y - nc)dy \end{bmatrix},$$

$$P_c : \begin{bmatrix} \varphi \\ \psi \end{bmatrix} (\xi) \mapsto \begin{bmatrix} \sum_{n=1}^{\infty} \eta_i f(\varphi(\xi - nc - c_i)) \\ \sum_{n=1}^{\infty} \rho_i f(\psi(\xi - nc - c_i)) \end{bmatrix},$$

and

$$A_c : \begin{bmatrix} \varphi \\ \psi \end{bmatrix} (\xi) \mapsto \begin{bmatrix} \beta\psi(x - nc - p_2) \\ \delta\psi(x - nc - q_2) \end{bmatrix}.$$

It is easy to see that a wave solution, that is, a solution to Eq.(1.1) with properties listed in Definition (2) is a solution to the equation

$$\phi = P_c\phi + Q_c\phi + A_c\phi.$$

Notice that, with our notations,

$$P_c\phi + Q_c\phi + A_c\phi = S_c[\phi] = S[\phi(\cdot + c)].$$

$$\phi = (I - A_c - P_c)^{-1} Q_c\phi.$$

From Definition (3), A_c and P_c are order preserving so, $(I - A_c - P_c)^{-1}$ is order preserving.

Let $M = (M_1, M_2)^T \in \mathbb{R}^2$ such that

$$M_1,\ M_2 > 0, \qquad S(M) = M.$$

Recall that C_M denotes the space of all continous function ϕ defined on $\mathbb{R}$ with values in $\mathbb{R}^2$ such that $0 \leq \phi \leq M$ for all x $\in \mathbb{R}$.

Lemma 4. *Let f and K satisfy the Standing Assumption, and*

$$(\beta + \sum_{i=1}^{N} \eta_i r) < 1,$$

$$(\delta + \sum_{i=1}^{N} \rho_i r) < 1.$$

Then, the operator $(I - A_c - P_c)^{-1}$ exists as a Lipschitz continuous operator from $BM(\mathbb{R}, [0, M_1] \times [0, M_2])$, and it leaves $BC(\mathbb{R}, [0, M_1] \times [0, M_2])$ invariant.

Proof. To show the existence of the operator $(I - A_c - P_c)^{-1}$ in $BM(\mathbb{R}, [0, M_1] \times [0, M_2])$, we will solve w $\in BM(\mathbb{R}, [0, M_1] \times [0, M_2])$ from the equation

$$w = A_c(w) + P_c(w) + u.$$

For any a fixed u $\in BC(\mathbb{R}, [0, M_1] \times [0, M_2])$, let

$$S_u(w) = A_c(w) + P_c(w) + u.$$

Then,

$$\|S(w) - S(v)\| = \|[A_c(w) + P_c(w)] - [A_c(v) + P_c(v)]\|.$$

By

$$[A_c(w) + P_c(w)] = \beta w(\xi - nc - p_2) + \sum_{n=1}^{\infty} \eta_i f(w(\xi - nc - c_i)),$$

$$[A_c(v) + P_c(v)] = \beta v(\xi - nc - p_2) + \sum_{n=1}^{\infty} \eta_i f(v(\xi - nc - c_i)),$$

we get

$$\begin{aligned}[A_c(w) + P_c(w)] - [A_c(v) + P_c(v)] &= [\beta w(\xi - nc - p_2) + \sum_{n=1}^{\infty} \eta_i f(w(\xi - nc - c_i))] \\ &\quad -[\beta v(\xi - nc - p_2) + \sum_{n=1}^{\infty} \eta_i f(v(\xi - nc - c_i))].\end{aligned}$$

Hence,

$$\begin{aligned}\|S(w) - S(v)\| &= \|[\beta w(t) + \sum_{n=1}^{\infty} \eta_i f(w(t))] - [\beta v(t) + \sum_{n=1}^{\infty} \eta_i f(v(t))]\| \\ &= \|\beta w(t) - \beta v(t) + \sum_{n=1}^{\infty} \eta_i f(w(t) - \sum_{n=1}^{\infty} \eta_i f(v(t))\| \\ &= \|\beta(w(t) - v(t) + \sum_{n=1}^{\infty} \eta_i [f(w(t)) - f(v(t))\| \\ &\leq \|\beta(w(t) - v(t)\| + \sum_{n=1}^{\infty} \eta_i \|[f(w(t)) - f(v(t))\| \\ &\leq \|\beta(w(t) - v(t)\| + \sum_{n=1}^{\infty} \eta_i sup\|[F'(t)\| * \|w - v\| \\ &= \|[\beta(w - v)\| + \sum_{n=1}^{\infty} \eta_i sup\|r/[1 + (r-1)t]^2\| * \|w - v\| \\ &= \|[\beta(w - v)\| + \sum_{n=1}^{\infty} \eta_i r\|w - v\| \\ &= (\beta + \sum_{n=1}^{\infty} \eta_i r)\|w - v\|. \\ &< \|w - v\|.\end{aligned}$$

So, S_u is a strictly contractive. By the Contraction Mapping Theorem, there is a unique fixed point w such that

$$w = \lim_{n \to -\infty} S_u^n(w),$$

$$w = (I - A_c - P_c)^{-1} u = \lim_{n \to -\infty} S_u^n(w).$$

□

Lemma 5. *Let f and K satisfy the Standing Assumption. Then, if ϕ is a bounded measurable function. Then, the function $Q_c(\phi)$ is a continuous function.*

Lemma 6. *Let all conditions of Lemma 4 be satisfied. Then, every solution $\phi \in BM(\mathbb{R}, \mathbb{R}^2)$ of the equation*

$$\phi = P_c\phi + Q_c\phi + A_c\phi.$$

is a continuous function.

2.2. Spreading Speed

According to [12] the spreading speed of the model (1.1) can be defined as follows: Let us choose a function $\varphi : \mathbb{R} \to [0, M]$ with the following properties

1. φ is continuous and non-increasing;
2. $\varphi(-\infty) := \lim_{t \to -\infty} \varphi(t) \in (0, M)$;
3. $\varphi(s) = 0$ for $s \geq 0$.

We then define an operator $R_c[\cdot]$ on the space C_M for every constant c as

$$R_c[u](s) := \max\{\varphi(s), Q[u(c + \cdot)](s)\}, \quad s \in \mathbb{R},$$

and a sequence of functions $\{a_n(c; \cdot)\}$ by

$$a_{n+1} := R_c[a_n], \quad a_0 = \varphi.$$

As shown in [12] the sequence $\{a_n(c;\cdot)\}$ is non-increasing and bounded, so for each $s \in \mathbb{R}$, we obtain the pointwise limit

$$\lim_{n\to\infty} a_n(c;s) = a(c;s), \quad s \in \mathbb{R}.$$

Obviously, $0 \le a(c;s) \le M$ for all $s \in \mathbb{R}$. The following number is called *spreading speed* for our model

$$c_+^* := \sup\{c | a(c;+\infty) = M\}.$$

Similarly we will define a number c_-^* as follows:

$$R_{c_-}[u](s) := \max\{\varphi(s), Q[u(c-\cdot)](s)\}, \quad s \in \mathbb{R},$$

and a sequence of functions $\{b_n(c;\cdot)\}$ by

$$b_{n+1} := R_{c_-}[b_n], \quad b_0 = \varphi.$$

Again, as shown in [12] the sequence $\{b_n(c;\cdot)\}$ is non-increasing and bounded, so for each $s \in \mathbb{R}$, we obtain the pointwise limit

$$\lim_{n\to\infty} b_n(c;s) = b(c;s), \quad s \in \mathbb{R}.$$

Obviously, $0 \le b(c;s) \le M$ for all $s \in \mathbb{R}$. Then,

$$c_-^* := \sup\{c | \, b(c;+\infty) = M\}.$$

By the theory in [12], if there exists a bounded non-negative measure $m(x, dx)$ on $\mathbb{R}$ such that

$$S[u](x) \le \int_{-\infty}^{\infty} u(x-y)m(y,\, dy), \quad u \in C_M,$$

then

$$\begin{aligned} c_+^* &\le \inf_{\mu>0} \frac{1}{\mu} \ln \int_{-\infty}^{\infty} e^{ux} m(x,\, dx), \\ c_-^* &\le \inf_{\mu>0} \frac{1}{\mu} \ln \int_{-\infty}^{\infty} e^{-\mu x} m(x,\, dx). \end{aligned}$$

And if there exists a bounded non-negative measure $l(x,\, dx)$ with property that

$$\int_{-\infty}^{\infty} l(x,\, dx) > 1, \quad S[u](x) \ge \int u(x-y)l(y,\, dy)$$

for all u such that $0 \leq u(x) \leq \epsilon$, then

$$c_+^* \geq \inf_{\mu>0} \frac{1}{\mu} \ln \int_{-\infty}^{\infty} e^{\mu x} l(x,\ dx),$$

$$c_-^* \geq \inf_{\mu>0} \frac{1}{\mu} \ln \int_{-\infty}^{\infty} e^{-\mu x} l(x,\ dx).$$

Clearly,

$$F(u) \leq r_0 u, \qquad u \in [0, r_0]$$

where $r_0 > 1$. Therefore, for each $u \in C_M$,

$$\begin{cases} u_{n+1} \leq \sum_{n=1}^{\infty} \eta_i r_0 u_n(x-c_i) + \alpha \int_{-\infty}^{\infty} K(x-p_i-y) r_0 u_n(y) dy + \beta w_n(x-p_2), \\ w_{n+1} \leq \sum_{n=1}^{\infty} \rho_i r_0 w_n(x-c_i) + \gamma \int_{-\infty}^{\infty} K(x-q_i-y) r_0 w_n(y) dy + \delta u_n(x-q_2), \end{cases}$$

$$m(x, dx) := (\beta\delta(x-p_2) + \sum_{i=1}^{\infty} \eta_i e^{r_0} \delta(x-c_i) + \alpha K(x-p_i) r_0) dx,$$

where $\delta(x)$ is the Dirac's delta function, that is, for all $a \in \mathbb{R}$,

$$\int_{-\infty}^{\infty} f(x)\delta(x-a)dx = f(a).$$

By the theory of spreading speed in [12], we have

$$\begin{aligned} c_+^* &\leq \inf_{\mu>0} \frac{1}{\mu} \ln \int_{-\infty}^{\infty} e^{\mu x} m(x,\ dx) \\ &= \inf_{\mu>0} \frac{1}{\mu} \ln \int_{-\infty}^{\infty} e^{\mu x} (\beta\delta(x-p_2) + \sum_{i=1}^{\infty} \eta_i e^{r_0} \delta(x-c_i) + \alpha K(x-p_i) r_0) dx \\ &= \inf_{\mu>0} \frac{1}{\mu} \ln \left(\beta e^{\mu p_2} + \sum_{i=1}^{\infty} \eta_i r_0 e^{\mu c_i} + \alpha r_0 \int_{-\infty}^{\infty} e^{\mu x} K(x-p_i) dx \right). \end{aligned}$$

Similarly

$$\begin{aligned} c_-^* &\leq \inf_{\mu>0} \frac{1}{\mu} \ln \int_{-\infty}^{\infty} e^{-\mu x} m(x,\ dx) \\ &= \inf_{\mu>0} \frac{1}{\mu} \ln \int_{-\infty}^{\infty} e^{-\mu x} (\beta\delta(x-p_2) + \sum_{i=1}^{\infty} \eta_i r_0 \delta(x-c_i) + \alpha K(x-p_i) r_0) dx \\ &= \inf_{\mu>0} \frac{1}{\mu} \ln \left(\beta e^{-\mu p_2} + \sum_{i=1}^{\infty} \eta_i e^{-\mu c_i} + \alpha r_0 \int_{-\infty}^{\infty} e^{-\mu x} K(x-p_i) r_0) dx \right). \end{aligned}$$

2.3. Existence of Traveling Waves

By using Monotone Iteration Method as shown in [5], we will construct a monotone sequence of functions that is convergent to a monotone traveling wave solution. We define a sequence of functions $\{\phi_n\}_{n=1}^{\infty}$ as follows: Set $\phi_1(s) := a(c; s)$. Since $a(c; \cdot)$ is non-increasing and bounded it is a measurable and bounded function on $\mathbb{R}$, that is, $a(c; \cdot) \in B_M$. Therefore, we can define

$$\phi_{n+1} = (I - A_c - P_c)^{-1} Q_c[\phi_n], \; n \geq 1, \quad n = 1, 2, \cdots.$$

The sequence $\{\phi_n(x)\}$ is convergent because it is non-increasing and bounded below by zero for every fixed $x \in \mathbb{R}$, Let's say it is convergent to $W(x)$.

The limit function $W(x)$ is a non-increasing function, so it is measurable. Moreover, by the Lebesgue's Dominated Convergence Theorem,

$$\begin{aligned}
\lim_{n\to\infty} A_c[\phi_n](x) &= \lim_{n\to\infty} [\beta\varphi_n(x - nc - p_2)] \\
&= [\beta W(x - nc - p_2)] \\
&= A_c[W](x).
\end{aligned}$$

$$\begin{aligned}
\lim_{n\to\infty} P_c[\phi_n](x) &= \lim_{n\to\infty} \sum_{n=1}^{\infty} \eta_i f(\phi_n i(\xi - nc - c_i)) \\
&= \sum_{n=1}^{\infty} \eta_i f(W(\xi - nc - c_i)) \\
&= P_c[W].
\end{aligned}$$

$$\begin{aligned}
\lim_{n\to\infty} Q_c[\phi_n](x) &= \alpha \int_{-\infty}^{\infty} K(\xi + nc + c - p_1 - y) f(\phi_n(y - nc) dy \\
&= \alpha \int_{-\infty}^{\infty} K(\xi + nc + c - p_1 - y) f(W(y - nc) dy \\
&= Q_c[W].
\end{aligned}$$

Therefore,

$$\begin{aligned}
W = \lim_{n\to\infty} \phi_{n+1}(x) &= \lim_{n\to\infty} S[\phi_n](x) \\
&= \lim_{n\to\infty} A_c[\phi_n](x) + \lim_{n\to\infty} P_c[\phi_n](x) + \lim_{n\to\infty} Q_c[\phi_n](x) \\
&= A_c[W](x) + P_c[W](x) + Q_c[W](x) \\
&= S_c[W](x).
\end{aligned}$$

Hence, W is a fixed point of Q_c in B_M. We have

$$W - A_c[W] - P_c[W] = Q_c[W],$$

and thus

$$W = (I - A_c - P_c)^{-1} Q_c[W],$$

Since W is continuous and is the limit of a sequence of monotone continuous functions, By Dini's Theorem, the convergence is uniform on each compact interval.

If

$$J_c[u] := (I - A_c - P_c)^{-1} Q_c[W]$$

in the function space $BM(\mathbb{R}, [0, M])$, then it has similar properties as S as listed in [4,lemma 2.3]

$$\begin{aligned} a(c; -\infty) &= M; \\ a(c; +\infty) &= 0. \end{aligned}$$

Since the sequence $\{\phi_n\}$ is convergent uniformly on each compact interval, by [12,Theorem 6.2] we have

$$\lim_{t \to -\infty} W(t) = M.$$

Moreover,

$$0 \leq W(+\infty) \leq \phi_1(+\infty) = 0.$$

So,

$$W(+\infty) = 0.$$

Thus, the boundary conditions for W are satisfied.

ACKNOWLEDGMENT

The author would like to thank Drs. Nguyen Van Minh, Jin Liang for their helpful discussions and encouragement.

REFERENCES

[1] Aronson D. G., Weinberger H. F., Multidimensional nonlinear diffusion arising in population genetics. *Adv. in Math.*, 30 (1978), no. 1, 33-76.

[2] Aronson D. G., Weinberger H., Nonlinear diffusion in population genetics, combustion, and nerve propagation. *Partial Differential Equations and Related Topics* (J. Golstein ed.). pp. 5-49. Lecture Notes in Mathematics, vol. 466. Springer. Berlin 1975.

[3] Lutscher F., Density-dependent dispersal in integrodifference equations. *J. Math. Biol.*, 56 (2008), 499-524.

[4] Lutscher F., Van Minh N., Monotone traveling waves in discrete models of biological populations with Sessile Stages.*Nonlinear Analysis: Real World Application.*, 14 (2013), 495-506.

[5] Le M., Van Minh N., Monotone Traveling Waves in a general discrete model for populations.*J. Discrete Cont Dyn.*, 22 (2017), 3221-3234.

[6] Li B., Traveling Wave solutions in a plant population modl with a seed bank. *J. Math. Bio.*, 65 (2012), 855-873.

[7] Hastings A., Cuddington K., Davies K. F., Dugaw C. J., Elmendorf A., Freestone A., Harrison S., Holland M., Lambrinos J., Malvadkar U., Melbourne B. A., Moore K., Taylor C., Thomson D., The spatial spread of invasions: new developments in theory and evidence. *Ecol. Lett.*, 8 (2005) 91-101.

[8] Li B., Lewis M. A., Weinberger H. F., Existence of traveling waves for integral recursions with nonmonotone growth functions. *J. Math. Bio.*, 58 (2009), 323-338.

[9] Li B., Lewis M. A., Weinberger H. F., Spreading speeds as slowest wave speeds for cooperative systems. *Math. Biosci.*, 196 (2005), 82-98.

[10] Weinberger H. F., Lewis M. A., Li B., Anomalous spreading speeds of cooperative recursion systems.*J. Math. Biol.*, 55 (2007), 207-222.

[11] Weinberger H. F., Asymptotic behavior of a model in population genetics. *Partial Differential Equations and Applications* (J. Chadam ed.). Lecture Notes in Mathematics, 648 (1978), 47–98.

[12] Weinberger H. F., Long-time behavior of a class of biological models. *SIAM J. Math. Anal.*, 13 (1982), 353-396.

In: Evolutionary Processes and Applications
Editor: G. M. N'Guérékata
ISBN: 978-1-53616-350-6

Chapter 2

Generation and Evolution of Magneto-Acoustic Waves in an Isothermal Atmosphere (I)

H. Y. Alkahby[1,*]***, D. A. Rossmanith***[1,†]***, R. A. Shulaiba***[2,‡]
and D. R. Jagessar[1,§]
[1]Department of Mathematics, Dillard University,
New Orleans, LA, US
[2]Department of Information Technology,
University of the Cumberlands, Williamsburg, KY, US

Abstract

In this chapter, we will study generation, propagation, reflection and transmission of magneto-acoustic waves in an isothermal solar atmosphere under the effect of arbitrary values of the effect of thermal conduction, viscosity and horizontal magnetic field. A general solution was obtained. It is shown that if the effect thermal conduction dominates the wave generation and propagation, the solar atmosphere can be divided into two different regions which are connected by an absorbing and reflecting layer. Below the reflecting layer, the oscillatory process is adiabatic and the solution can be written as a linear combination of upward

*Corresponding Author's Email: halkahby@dillard.edu.
†Corresponding Author's Email: drossmanith@dillard.edu.
‡Corresponding Author's Email: refaat.shulaiba@ucumberlands.edu.
§Corresponding Author's Email: daniel.jagessar@dillard.edu.

and down ward propagating waves. Above the reflecting layer, the solution decays exponentially with the altitude. Also, if the combined effects of viscosity and magnetic field dominate the wave's generation, the atmosphere can be divided into two different regions which are connected by an absorbing and reflecting layer and the effect of thermal conduction can be eliminated. In both cases the reflection coefficient was determined. The general solution was compared to another solution in which the inner and the outer approximate solutions were matched in an overlapping domain. The results of the two solutions were investigated in connection with the heating mechanisms of the solar atmosphere.

Keywords: magneto-acoustic waves, reflecting layer, magneto-viscous diffusivity, thermal diffusivity

1. Introduction

Generation, reflection and transmission of magneto-acoustic waves in an isothermal atmosphere has been studied extensively in recent years. The motivation for these studies stems from their application to phenomena in compressible ionized fluids, such as solar, stellar, earth atmosphere and possible other applications [1, 2, 3, 4, 5, 6, 7, 8, 10, 13, 30 and the references there in]. The presence of dissipation in an isothermal atmosphere may result in upward traveling, small amplitude waves to be reflected downward. This type of reflection is more significant when the wavelength is largely compared to the density scale height [8, 10, 23, 26, 28, 29].

In this chapter, we investigate the combined effects of viscosity, thermal conduction and a horizontal magnetic field on vertically propagating magneto-acoustic waves in an isothermal atmosphere for arbitrary values of magneto-viscous and thermal diffusivities. This problem leads to a singular perturbation problem with three terms recursion formula and it is solved by Laplace Transform when the effect of thermal conductivity is large compared to the combined effects of the kinematic viscosity and the Alfvén speed. It is shown that the atmosphere can be divided into two regions which are connected by a transition region in which the reflection and the wave modification takes place. In the lower region, the effects of the thermal diffusivity, kinematic viscosity and Alfvén speed are negligible, while in the upper region, the effect of these quantities is large because of the exponential decrease of the density with altitude. As a result, in the lower region, the motion is adiabatic and in the upper region, it is isothermal because the exponential increase of the thermal diffusivity with altitude transforms oscillatory notion from an adiabatic form, below the reflecting, to an isothermal one above it . Consequently, the upward propagating waves in the adiabatic region will be reflected downward by the transition region. Also it is shown that if the combined effect of the kinematic viscosity and Alfvén speed is dominant, the effect of thermal conduction can be eliminated and the atmosphere can be, again, divided in two distinct regions which are connected by an absorbing and reflecting layer. In this case, the nature of the reflecting layer depends on the relative strength of the kinematic viscosity with respect to that of Alfvén speed. If the effect of the viscosity is dominant, the reflecting layer acts like an absorbing and reflecting layer. At the same time if Alfvén speed is dominant the reflecting layer act like a totally reflecting barrier because of the magnetic field does not generate dissipation mechanism. Thus, the maximum value of the reflection coefficient depends on the relative strength of the kinematic viscosity and Alfvén speed with respect to thermal diffusivity. The solution by Laplace Transform is compared to an approximate solution in which the inner and outer approximations are matched in an overlapping domain [10, 2]. In all cases the reflection coefficient is determined and the conclusions of both solutions are investigated in connection with the heating mechanisms of the solar atmosphere.

2. Mathematical Formulation of the Problem

Suppose an isothermal atmosphere, which is viscous and thermally conducting, and occupies the upper half-plane, $z > 0$. We will investigate the problem of small vertical oscillations about equilibrium, i.e. oscillations which depend only on time t, and on the vertical coordinate z. Let the equilibrium pressure, density, magnetic field, and temperature be denoted by P_0, ρ_0, B_0, and T_0, where P_0 and T_0 satisfy the gas law $P_0 = R\,T_0\,\rho_0$ and the hydrostatic equation $P_0' + g\,\rho_0 = 0$. Here, R is the gas constant, g is the gravitational acceleration, and the prime denotes differentiation with respect to z. The equilibrium pressure and density are given by:

$$P_0(z) = P_0(0)\exp\big(-\frac{z}{H}\big)\,,\ \rho_0(z) = \rho_0(0)\exp\big(-\frac{z}{H}\big)\,,$$

where $H = R\,T_0/g$ is the density sacale and height.

Let p, ρ, ω. B, and T be the perturbations in the pressure, density, vertical velocity, magnetic field, and temperature. The linearized equations of motion are:

$$\omega_t + p_z + g\,\rho = \frac{4}{3}\mu\,\omega_{zz}, \tag{1}$$

$$\rho_t + (\rho_0\,\omega)_z = 0, \tag{2}$$

$$B_t + (B_0\,\omega)_z = 0, \tag{3}$$

$$\rho(c_\nu(T_t + T) + g\,H\,\omega_z) = \kappa\,T_{zz}, \tag{4}$$

$$p = R(\rho_o T + T_0\,\rho). \tag{5}$$

These are, respectively, the equations for the change in vertical momentum, the mass conservation equation, the heat flow equation, and the gas law.Here, c_ν is the specific heat at constant volume, μ is the dynamic viscosity coefficient, and $kappa$ is the thermal conductivity, all assumed to be constants. The subscript z and t denote differentiation with repsect to z and t respectively. We will consider solutions which are harmonic in time, i.e:

$$\omega(z,t) = W(z)\exp(-i\,\sigma\,t), \tag{6}$$

$$T(z,t) = T(z)\exp(-i\,\sigma\,t), \tag{7}$$

where σ denotes the frequency of the wave.

It is more convenient to rewrite the equations in a dimensionless form:

$$\acute{z} = \frac{z}{H} \qquad \acute{t} = \frac{t}{\sigma_\alpha} \qquad \acute{T} = \frac{T}{2\gamma T_0}$$

$$sig\acute{m}a = \frac{\sigma}{\sigma_a} \qquad \sigma_a = \frac{c}{2\,H} \qquad \acute{\kappa} = \frac{2\,\kappa}{c_\nu\,c\,H\,\rho_0(0)}$$

$$\acute{\mu} = \frac{2\,\mu}{3\rho(0)c\,H} \qquad \alpha = \frac{c^2\,A}{c^2} \qquad \acute{\eta} = \alpha - i\,\acute{\eta}$$

where $c = \sqrt{\gamma\,R\,T_0} = \sqrt{\gamma\,g\,H}$ is the adiabatic sound speed and σ_a is the adiabatic cutoff frequency. The primes can be omitted since all variables will be written in dimensionless form from now on.

One can eliminate p, B, and ρ from equation (1) by differentiating the equation with respect to t and substituting equations (2)-(5) to obtain a system of differential equations for $W(z)$ and $T(z)$,

$$\left(D^2 + D + \frac{\gamma\sigma^2}{4}\right)W(z) + \gamma\mu e^z D^2 W(z) + i\gamma(D-1)T(z) = 0, \quad (8)$$

$$(\gamma - 1)DW(z) = \gamma(i\sigma)T(z) + \gamma\kappa e^z D^2 T(z), \quad (9)$$

where $D = d/dz$. Furthermore, if $W(z)$ can be eliminated from the differential equation (8) by differentiating with respect to z and substituting for $DW(z)$ from differential equation (9), one obtains a single, fourth-order equation for $T(z)$:

$$\begin{aligned}&\Big(\gamma\sigma\big(D^2 - D + \frac{\sigma^2}{4}\big) - i\kappa e^z D^2\big(D^2 + D + \frac{\gamma\sigma}{4}\big)\\&+ \gamma\sigma\eta\kappa e^z D(D+1) - i\gamma\eta\kappa e^{2z} D^2(D+1)(D+2)\Big)T(z) = 0,\end{aligned} \quad (10)$$

In addition, it is convenient to introduce the dimensionless parameter, $\epsilon = \eta/\kappa$, which measures the relative strength of the combined effect of the viscosity and the magnetic field with respect to thermal conductivity. Consequently, equation (10) becomes:

$$\begin{aligned}&\Big(\gamma\sigma\big(D^2 - D + \frac{\sigma^2}{4}\big) - i\kappa e^z D^2\big(D^2 + D + \frac{\gamma\sigma}{4}\big)\\&+ \epsilon\gamma\sigma\kappa e^z iD(D+1) - i\gamma\epsilon(\kappa e^z)D^2(D+1)(D+2)\Big)T(z) = 0.\end{aligned} \quad (11)$$

2.1. Boundary Conditions

Physically relevant solutions must also satisfy the following conditions:

$$\mu\int_0^\infty |W_z|^2 dz < \infty, \quad (1)$$

$$\kappa\int_0^\infty |T_z|^2 dz < \infty. \quad (2)$$

The first of these equations is the dissipation condition (DC), which follows from the finiteness of the energy dissipation rate in an infinite column of fluid of a unit cross-section per period of oscillation. The second one, the entropy condition (EC), is a consequence of the finiteness of the entropy growth rate in a column of fluid. These conditions are necessary and sufficient (as upper boundary conditions) to determine a unique solution. The upper boundary conditions are inapplicable if $\kappa = \mu = 0$, and, for uniqueness, we will impose the radiation condition (see Alkahby, 1994). Boundary conditions are also required at $z = 0$, and we shall adopt the lower boundary condition (LBC): In a fixed interval $0 < z < z_0$, the solution of the differential equation (11) should approach some solutions of the limiting differential equations ($\kappa \to 0$ and $\eta \to 0$),

$$\left(D^2 + D + \frac{\sigma^2}{4}\right) T(z) = 0. \tag{3}$$

Considering the lower boundary condition is simpler than prescribing $T(z)$, $W(z)$ at $z = 0$.

3. Heuristic Discussion of the Problem

At the outset, we have to indicate that we are considering the problem when $|\epsilon|$ is sufficiently small, and the dimensionless parameters α, μ, and κ are sufficiently small and proportional to the values at $z = 0$ of the Alfvén speed, kinematic viscosity, and thermal diffusivity. Also, we are assuming that the solution of the differential equation (11) is uniform with respect to the parameters η and κ as these parameters tend to zero. We call $|\eta|/\rho_0$ the magneto-viscous diffusivity. Then, the parameter $|\epsilon$ can be written in the form:

$$|\epsilon| = |\eta/\kappa| = |\frac{(\eta/\rho_0)}{(\kappa\, \rho_0)}|. \tag{1}$$

As a result, the parameter $|\epsilon|$ measures the relative strength of the thermal diffusivity, κe^z, with respect to the magneto-viscous diffusivity $|\eta| e^z$. Since we are investigating the case of small $|\eta|$, we have $\kappa e^z << |\eta| e^z$. Consequently, for $|\eta| e^z << 1$, we have three distinct regions which are connected by two different reflecting layers, on which the reflection, dissipation, and wave modification take place. In the lower region, $0 < z < z_0 = -\log(\eta)$, solutions of the differential equation (11), below the reflecting layer, can be approximated by solutions of the following differential equation:

$$\left(D^2 - D + \frac{\sigma}{4}\right) T(z) = 0. \tag{2}$$

Furthermore, in the lower region, the effect of the magneto-viscous diffusivity and the thermal diffusivity is negligible. In the upper region, $z_1 = -\log\kappa < z << z_2 = -\log|\alpha|$, where $|\eta|e^z >> 1$, the solutions of the differential equation (1) can be approximated by the solutions of the following differential equation:

$$\Big(D^2(D+1)(D+2)\Big)T(z) = 0. \tag{3}$$

Solutions of this differential equation, which satisfy the boundary conditions (1) and (2), must behave either as constants or decay exponentially with altitude z. In the upper region, solutions of the differential equation (3) will be influenced by the combined effects of the viscosity, thermal conduction, and the magnetic field. In the middle region, the solution will be approximated by the solution of the following equation:

$$\left(D^2 - D + \frac{\gamma\sigma^2}{4}\right)T(z) = 0. \tag{4}$$

The upper and middle regions are connected by a transition region, in which the reflection of the waves and dissipation of the wave energy take place. Moreover, in the transition region, magneto-viscous diffusivity values change from small to large. When the magneto-viscous diffusivity is large compared to the thermal diffusivity, the solution of the differential equation (11) can be approximated by the solution of the following differential equation:

$$\left((D^2 - D + \frac{\gamma\sigma^2}{4}) + \gamma\eta e^z(D^2 + D)\right)T(z) = 0. \tag{5}$$

4. Series and Asymptotic Solutions of the Problem

To obtain the solution to equation (11), it is convenient to introduce a new dimensionless variable χ, defined by

$$\chi = \frac{exp(-z)}{i\kappa} = exp\big(-z - \log\kappa + \frac{3\pi i}{2}\big), \tag{1}$$

which transforms the differential equation (11) into:

$$\begin{aligned}&\Big(\gamma\sigma\chi^2(\phi^2+\phi+\frac{\sigma^2}{4})-\chi\phi^2(\phi^2-\phi+\frac{\gamma\sigma^2}{4})\\&-i\epsilon\gamma\sigma\chi(\phi^2-\phi)+\gamma\sigma\phi^2(\phi-1)(\phi-2)\Big)T(\chi)=0.\end{aligned} \tag{2}$$

where $\phi = \chi\frac{d}{dx}$. It is clear that the point $\chi = 0$ corresponds to $z \to \infty$, the point $\chi_0 = \exp\left(9 - \log\kappa + (3\pi i)/2\right)$ to $z = 0$ and the segment connecting these points in the complex χ plane to $z \geq 0$. As $\kappa \to 0$, the point $\chi_0 \to \infty$.

The point $\chi = 0$ is a regular singular point of equation (2). Consequently, there are four linearly independent solutions in the neighborhood of $\chi = 0$, which for arbitrary values of η, can be written in the following form:

$$T_1(\chi) = \sum_{n=0}^{\infty} a_n(e_1)\chi^{n+e_1}, \tag{3}$$

$$T_2(\chi) = \sum_{n=0}^{\infty} a'_n(e_2)\chi^{n+e_2} + T_1(\chi)\log\chi, \tag{4}$$

$$T_3(\chi) = \sum_{n=0}^{\infty} a''_n(e_3)\chi^{n+e_3}, \tag{5}$$

$$T_4(\chi) = \sum_{n=0}^{\infty} a'''_n(e_4)\chi^{n+e_4} + 3T_3(\chi)\log\chi, \tag{6}$$

where $e_1 = 2$, $e_2 = 1$, $e_3 = e_4 = 0$. The prime denotes differentiation of a_n with respect ot its argument. The coefficients $a_n(e_i)$ are determined from the following three term recursion formula:

$$p_0(n+2+e)a_{n+2} + p_1(n+1+e)a_{n+1} + p_2(n+2)a_n = 0, \tag{7}$$

where:

$$\begin{aligned}p_0(y) &= \gamma\eta y^2(y-1)(y-2),\\ p_1(y) &= -\eta y(y-1) - y^2(y^2 - y + \frac{\gamma\sigma^2}{4}),\\ p_2(y) &= \gamma\sigma(y^2 + y + \frac{\sigma^2}{4}).\end{aligned}$$

To determine which of the solutions defined in equations (3) - (6), satisfies the boundary conditions of equation (2) for large z, the solutions must be transformed to the variable z by means of equation (1). Thus, for $\kappa > 0$ and $P_r > 0$, we have:

$$T_1(z) = O(e^{-2z}), \tag{8}$$
$$T_2(z) = O(e^{-z}), \tag{9}$$
$$T_3(z) = O(1), \tag{10}$$
$$T_4(z) = O(z). \tag{11}$$

It is clear that $TY_4(z)$ is the only solution which does not satisfy the entropy condition.

To apply the dissipation condition, equation (8) must be used to determine the amplitudes of the velocity corresponding to the solutions defined in equations (3) - (6). As a result, we have:

$$DW_1(z) = O(e^{-z}), \tag{12}$$
$$DW_2(z) = O(1), \tag{13}$$
$$DW_3(z) = O(e^{-z}), \tag{14}$$
$$DW_4(z) = O(z). \tag{15}$$

To determine the linear combination of $T(\chi)$ in equations (12) - (15), the behavior of $T_1(z)$ and $T_3(z)$ for small z must be found. Recall that small z corresponds to large $|\chi|$ with $\arg(\chi) - 3\pi/2$. Thus, the asymptotic expansions of $T_1(\chi)$ and $T_3(\chi)$ about infinity should be found.

The point $\chi = \infty$ is an irregular singular point of the differential equation (2), and there are four linearly independent solutions whose asymptotic behavior, to first order, is governed by:

$$T_1^*(\chi) \sim \chi^{\alpha_1}(1 + h_{11}\chi^{-1} + h_{12}\chi^{-2} + ...), \tag{16}$$
$$T_2^*(\chi) \sim \chi^{\alpha_2}(1 + h_{12}\chi^{-1} + h_{12}\chi^{-2} + ...), \tag{17}$$
$$T_3^*(\chi) \sim \chi^{-\frac{1}{4}}(1 + h_{31}\chi^{-\frac{1}{2}} + ...) \exp(-s\chi^{\frac{1}{2}}), \tag{18}$$
$$T_4^*(\chi) \sim \chi^{-\frac{1}{4}}(1 + h_{41}\chi^{-\frac{1}{2}} + ...) \exp(-s\chi^{\frac{1}{2}}), \tag{19}$$

where

$$\alpha_1 = -\frac{1}{2} + i\beta,$$
$$\alpha_2 = -\frac{1}{2} - i\beta,$$
$$2\beta = \sqrt{\sigma^2 - 1},$$
$$s = 2\sqrt{\gamma\sigma}.$$

Reintroducing the dimensionless variable z by means of equation (1), we have:

$$T_1^*(z) \sim \exp\left((\frac{1}{2} - i\beta)z \right), \tag{20}$$

$$T_2^*(z) \sim \exp\left((\frac{1}{2} + i\beta)z \right), \tag{21}$$

$$T_3^*(z) \sim \exp\left(-(\frac{\gamma\sigma}{2\kappa})^{\frac{1}{2}} z \right), \tag{22}$$

$$T_4^*(z) \sim \exp\left((\frac{\gamma\sigma}{2\kappa})^{\frac{1}{2}} z \right). \tag{23}$$

(24)

As a result, $T_1^*(z)$ represents a downward propagating wave, while $T_2^*(z)$ is an upward propagating wave. Also, $T_3^*(z)$ corresponds to the boundary layer term. It is important only near $z = 0$, for any small fixed value of (z), and it decays with (z) like $\exp(-(\frac{\gamma\sigma}{2\kappa})^{1/2}z)$. Thus, it is more convenient to use the lower boundary condition from which it follows that the solution of the differential equation (2), when $\kappa \to 0$, should behave asymptotically like a linear combination of $T_1^*(\chi)$ and $T_2^*(\chi)$.

5. Magnitude of the Reflection Coefficient and Conclusions

It has been indicated in section 4 (equation (12) - (15)), that the solution of the boundary value problem, which satisfies the prescribed boundary condition, can be written as:

$$T(\chi) = c_1 T_1(\chi) = c_3 T_3(\chi), \tag{1}$$

where c_1 and c_3 are constants and can be determined such that the asymptotic expansion of $T(\chi)$ does not increse to infinity. To do this, we have to investigate the existence of the solutions of the recursion formula. That is, we try to determine a function $u(x)$ of the complex variable x such that, for any integer n, $u(n) = h(x+e) = h(y)$ and equation (7) becomes:

$$f_0(y)h(y+2) + f_1(y)h(y+1) + f_2(y)h(y) = 0, \tag{2}$$

where:

$$\begin{aligned} f_0 &= p_0(y+2), \\ f_1(y) &= p_1(y+1), \\ f_2(y) &= p_2(y). \end{aligned}$$

Following the same procedure as in Ford [5], two linearly independent solutions, $h_1(y)$ and $h_2(y)$ of the difference equation (1) may be determined as follows:

$$h_1(t) = t_1^y \int_1 t^y \Psi_1(t)dt, \tag{3}$$

$$h_2(t) = \frac{t_2^y}{\Gamma^2(y+1)} \int_1 t^y \Psi_2(t)dt. \tag{4}$$

The constants t_1 and t_2 are determined such that the leading coefficients of the two polynomials of the largest degree have equal absolute values. As a result, $t_1 = \eta\sigma$ and $t_2 = \gamma\sigma$. We define $h_1(y)$ as follows:

$$h_1(y) = t_1^y J_1(y). \tag{5}$$

The Laplace integral can be used to determine $J_1(y)$:

$$J_1(y) = \frac{1}{2\pi i} \int_1 t^{y-1} \Psi_1(t)dt. \tag{6}$$

Substituting equations (1) and (2) into the recursion formula (20)-(21), we have:

$$f_0(y)J_1(y+2) + \frac{f_1(y)}{t_1} J_1(y+1) + \frac{f_2(y)}{t_1^2} J_1(y) =$$

$$\frac{1}{2\pi i} \int_1 t^{y-1} \Big(t^4\theta_4(t)D^4 - t^3\theta_3(t)D^3 - t^2\theta_2(t)D^2 - t\theta_1(t)D + \theta_0(t) \Big) \Psi_1(t)$$

$$+ \frac{1}{2\pi i} I_1[(\Psi, t)]_l = 0, \tag{7}$$

where:

$$\theta_4(t) = t^2 - t,$$
$$\theta_3(t) = -3t^2 + 7t,$$
$$\theta_2(t) = 6t^2 + (2 + 2P_r\sigma^2 + \frac{\gamma\sigma}{4})t + \gamma\sigma^2,$$
$$\theta_1(t) = -t^2 - 2t,$$
$$\theta_0(t) = \frac{\gamma\sigma^2}{4}.$$

By considering the solution of the following differential equation:

$$\Big(t^4\theta_4(t)D^4 - t^3\theta_3(t)D^3 - t^2\theta_2(t)D^2 - t\theta_1(t)D + \theta_0(t)\Big)\Psi_1(t) = 0, \quad (8)$$

at $t = 0$ and $t = 1$, and observing the behavior of $I_1[(\Psi, t)]$, it is found that:

$$\Psi_1(t) = (t-1)\big(1 + h_{3,1}(t-1) + ...\big) + (t-1)\log 9t - 1)\big(h_{3,0} + ...\big). \quad (9)$$

The line integration l is the branch cut for $\log(t-1)$ and $\log(t)$. However, this is only valid for y such that real$(y) > 5/4$. In the same way, one can find the second solution $h_2(t)$. As a result, $J_1(y)$ is obtained by substituting $\Psi_1(t)$ into equation (1).

The second step in the analysis is to use the linear combination of $h_1(y)$ and $h_2(y)$ in determining the asymptotic expansion of $T_1(\chi)$ and $T_3(\chi)$. To do this, let $H_{12}(y) = \alpha h_1(y) + bh_2(y)$ such that $H_{12}(n+e) = a_n(e_1)$, where the constants α and b can be determined from $\alpha h_1(1) + bh_2(1) = 0$ and $\alpha h_1(2) + bh_2(2) = 1$. It follows that the asymptotic expansion of $T_1(\chi)$ can be written in the following form:

$$T_1(\chi) = \sum_{n=0}^{\infty} a_n\chi^{n+e_1} = \sum_{n=0}^{\infty} h_{12}(n+e)\chi^{n+e_1}$$
$$\sim -\sum_{poles}\left(\frac{\pi h_{12}(w+e_1)(-1)^w\chi^{w+e_1}}{\sin(\pi w)}\right) + \left(\frac{b\exp\left(i\pi(2e_1 + \frac{\pi}{2})\right)}{2\sqrt{\pi t_2^{\frac{1}{4}}}}\right)T_3^*\chi. \quad (10)$$

The sum is taken over all of the simplest poles $h_{12})$ at $w = \alpha_1 - n$ and at $w = \alpha_2 - n$, where $n = 1, 2, ...$ In the same way, the asymptotic expansion of

$T_3(\chi)$ can be written as:

$$T_3(\chi) \sim -\sum\left(\frac{h_{13}(w)(-\chi)^w}{\sin(\pi w)} + \left(\frac{id}{2\sqrt{\pi t_2^{\frac{1}{4}}}}\right)T_3^*\chi, \tag{11}$$

where $h_{13}(y) = ch_1(y) + dh_2(y) = a''_N(e_3)$. The constants c and d are determined from$ch_1(1) + dh_2(1) = \sigma$, and $ch_1(2) + dh_2(2) = 0$. Substitution of equation (19) - (21) into equations (5)-(6), and application of the condition that the coefficient of $T_3^*(\chi$ must be zero, we have the following form for the solution of the boundary value problem, defined by the differential equation (11), for arbitrary values of thermal conduction, viscosity, and magnetic field:

$$T(z) \sim -Const.\Big(\exp\big((\frac{1}{2} + i\beta)z\big)\Big) + K_\alpha \exp\Big((\frac{1}{2} - i\beta)z\Big), \tag{12}$$

where K_α denotes the reflection coefficient, defined by:

$$K_\alpha = \exp\Big(-\pi\beta - 2i\beta(\log\kappa)\Big)\frac{[h_1(y)]_{y=\alpha_1}}{[h_2(y)]_{y=\alpha_2}}. \tag{13}$$

It is clear that

$$|K_\alpha| = \exp(-\pi\beta)\frac{[h_1(y)]_{y=\alpha_1}}{[h_2(y)]_{y=\alpha_2}}. \tag{14}$$

Moreover, the forms of $h_1(y)$ and $h_2(y)$ are very complicated for the determination of the residues. At this point, we will determine the magnitude of the reflection coefficient by perturbation expansion, i.e.:

$$K_\alpha = K_{\alpha,0} + \mathcal{O}(\eta), \tag{15}$$

which can be obtained from the perturbation expansion of:

$$h_1(y) = h_{1,0}(y) + \eta h_{1,1}(y) + \mathcal{O}(\eta^2), \tag{16}$$

$$h_2(y) = h_{2,0}(y) + \eta h_{2,1}(y) + \mathcal{O}(\eta^2). \tag{17}$$

Substituting these expansions into the recursion formula (7), results in equation (1) by setting $\eta = 0$. The resulting form of the reflection coefficient is:

$$K_\alpha \sim K_{\alpha,0} = \Omega\exp(-\pi\beta - 2i\beta(\log(\gamma\sigma\kappa))), \tag{18}$$

where,

$$\Omega = \frac{\Gamma(\alpha_1 - \alpha_2)\Gamma^2(\alpha_2 + 1)\Gamma(\alpha_2 - \beta_1)\Gamma(\alpha_1 - \beta_2)}{\Gamma(\alpha_2 - \alpha_1)\Gamma^2(\alpha_1 + 1)\Gamma(\alpha_1 - \beta_1)\Gamma(\alpha_2 - \beta_2)}. \tag{19}$$

Here, $\beta_1 = -\frac{1}{2} - i\beta^*$, $\beta_2 = -\frac{1}{2} - i\beta^*$, and $2\beta^* = \sqrt{\gamma\sigma^2 - 1}$.

Finally, it is clear that, for $\beta > 0$,

$$|K_\alpha| = \exp(-\pi\beta) \leq 1. \tag{20}$$

6. Solution of Equation (2) by Matching Inner and Outer Approximations in an Overlapping Domain

In this section, we will find and investigate the behavior of the solution to equation (2). It is clear that equation (9) leads to a singular perturbation problem, and the solution can be found by matching the inner and outer approximations in an overlapping domain. This method will reduce the three term recursion difference equation to a two term recursion equation. The differential equation for the inner approximation $I(z)$ can be obtained by letting $\eta = 0$ in equation (2). The resulting equation is:

$$\Bigg(\gamma\sigma(D^2 - D + \frac{\sigma^2}{4}) - i\kappa e^z D^2(D^2 + D + \frac{\gamma\sigma^2}{4}) bigg) I(\chi) = 0. \tag{1}$$

This equation represents the effect of thermal conductivity only, and the solution can be written as in [5]:

$$T(z) = \exp\left((\frac{1}{2} + I\beta)z\right) + K_R \exp\left((\frac{1}{2} - I\beta)z\right), \tag{2}$$

where:

$$|K_R| = e^{\pi\beta}\left(\frac{\sinh(\pi(\beta_i - \beta))}{\sinh(\pi(\beta_i + \beta))} < e^{-\pi\beta}. \tag{3}$$

The differential equation of the outer approximation is obtained by introducing the stretched variable $\xi = \chi\alpha$. Letting $T(\xi) = U(\chi) + \mathcal{O}(\eta^2)$, and retaining the most significant terms in equation (9), we obtain:

$$\left(\xi(D^2 - D + \frac{\gamma\sigma^2}{4}) - i\gamma(D-1)(D-2)\right)D^2U(\xi) = 0, \tag{4}$$

where $D = \xi\frac{d}{d\xi}$. Since $\xi \to \infty$ for every fixed value of χ as $|\alpha| \to 0$, the asymptotic behavior of $I(\chi)$ as $\chi \to 0$ should be matched with $U(\xi)$ as $\xi \to \infty$.

Equations (8) and (9) are, respectively, similar to equations equations (4.6) and (4.9) in [19]. We will omit them and merely indicate the results.

The solution in the upper region will decay exponentially with altitude to a constant as $z \to \infty$, while in the two lower regions, it is a linear combination of upward and downward propagating waves. In the middle (isothermal) region, for $\sigma > \gamma^{-\frac{1}{2}}$, the solution can be written in the form:

$$T_i(z) \sim Const.\Big(\exp\left((-\frac{1}{2} + i\beta_i)z\right) - RC_i \exp\left((\frac{1}{2} - i\beta_i)z\right), \tag{5}$$

$$RC_i = \exp\left(2\beta_i\omega_1 + 2i(\omega - \beta_i \log|\gamma\eta|)\right), \tag{6}$$

$$\beta_i = \frac{1}{2}\sqrt{\gamma\sigma^2 - 1}, \tag{7}$$

$$\omega_1 = \arg(\eta) = \arg(\alpha), \tag{8}$$

$$\omega = \arg\left((\frac{1}{2} - i\beta_i)\Gamma^2(\frac{3}{2} - i\beta_i) \right). \tag{9}$$

It is clear that the maximum value of $|RC_i| = \exp(2\beta_i\omega_1)$. Thus, if the effect of the viscosity dominates the effect of the magnetic field, then $\omega_1 = \arg(\eta) = -\frac{\pi}{2}$, $|RC_i| = \exp(-\pi\beta_i)$, and the reflecting layer acts like an absorbing and reflecting barrier. In addition, if the effect of the magnetic field dominates the effect of the viscosity, then $\omega_1 = \arg(\eta) = 0$, $|RC_i| = 1$, and the reflecting layer acts laike a totally reflecting barrier. This result is in accordance with the nature of the magnetic field, because it does not procuce dissipation mechanisms.

In the lower (adiabatic) region, for $\sigma > 1$, the solution has the following form:

$$T_a(z) \sim Const.\Big(\exp\left((-\frac{1}{2} + i\beta_a)z\right) + RC_a \exp\left((\frac{1}{2} - i\beta_a)z\right), \tag{10}$$

$$RC_a = \exp(-\pi\beta_a)\exp\big(2i(\theta_2 - \beta_i \log(\gamma\sigma\kappa))\big), \tag{11}$$

$$\beta_a = \frac{1}{2}\sqrt{\sigma^2 - 1}, \tag{12}$$

$$\theta_2 = \arg\frac{\Gamma^2(\frac{1}{2} - i\beta_a\Gamma(2i\beta_a))\Gamma(i\beta_i - i\beta_a)}{\Gamma(i\beta_i + i\beta_a)}, \tag{13}$$

$$\phi_1 = \exp(2\pi\beta_i) - \exp(2\pi\beta_a), \tag{14}$$

$$R_a = \frac{\phi_1 - \phi_2 \exp A}{\phi_3 - \phi_4 \exp A}, \tag{15}$$

$$\phi_1 + \phi_2 = \phi_3 + \phi_4 = 2\sinh(2\pi\beta_i), \tag{16}$$

$$\phi_3 - \phi_1 = \phi_4 - \phi_2 = 2\sinh(2\pi\beta_a), \tag{17}$$

$$A = \beta_i(\pi + 2\phi_1 + 2i(\phi_3 - \beta_i \log|\alpha|)), \tag{18}$$

$$\phi_3 = \arg\frac{(-\frac{1}{2} + i\beta_i)\Gamma^2(2i\beta_i)}{\Gamma(i\beta_i + \beta_a)\Gamma(i\beta_i - i\beta_a)}. \tag{19}$$

7. Solution Behavior and Conclusion

From the above results, we have the following conclusions:

1. Equation (10) indicates the atmosphere can be divided into two distinct regions, upper and lower, connected by a reflecting and absorbing transision region where the wave reflection and modification takes place. In this case, the reflection resulted from the accumulative effect of thermal and magneto-viscous diffusivities. In the lower region, the oscillatory process can be represented by a linear combination of an upward and downward propagating wave. In the upper region, the solutions that satisfy the upper boundary conditions must decay with altitude. In the lower region, the motion of the waves will continue until their energy has dissipated, contributing to the heating mechanism of the solar atmosphere according to the wave theory for heating of the solar atmosphere.

2. Equation (1) represents the solution of the differential equation for the effect of thermal conduction only (equation (20)) and it indicates that the atmosphere can be divided into two regions which are connected by a partially reflecting layer. Since the upper boundary condition cannot be applied on the solution of the linear approximation, part of the wave continues to proagate upward. Once again, the solution in the lower region can be represented as a linear combination of an upward and a downaward propagating wave, and the solution will continue until the energy is dissipated in the solar atmosphere.

3. Equation (4) represents the solution of the differential equation (11) for the effect of thermal conduction, viscosity, and the magnetic field. In this case, the atmosphere can be divided into three distinct regions. The lower region is adiabatic, while the middle and upper regions are isothermal as a result of the relative diffusivity strengths. The lower two regions

are connected by a partially reflecting layer. because of the existence of two reflecting layers, the behavior of the solution can be described as follows: in the lower region, the upward propagating waves, incident waves, will be partially reflected by the first reflecting layer, which is created by thermal conduction, and the other part of the wave will be transmitted to the middle region. In the middle region, the motion will continue upwardsuntil it reaches the second reflecting layer, which is an absorbing and reflecting layer. Once again, at this layer, part of the wave will be reflected downward toward the partially reflecting layer, becoming incident waves, while the other part of the wave is dissipated at the upper reflecting layer. When the downward propagating waves reach the lower reflecting layer, part of the wave will be reflected upward, while the rest will be transmitted to the lower region. The process will continue until the waves deposit all of their energies, which may contribute to the heating process of the solar atmosphere.

4. It is clear from the above discussion that we have two transition regions in which the reflection, absorption, and wave modification take place. The first transition region, created by thermal conduction, acts like a partially reflecting layer. The second behaves like an aborbing and reflecting layer. The partially reflecting layer modifies the wave, and changes its wavelength from the adiabatic one in the lower region, to an isothermal one in the middle region for the upward propagating waves and from an isothermal one to an adiabatic one for the downward propagating waves. In addition, the absorbing and reflecting layer nature depends on the relative strength of the kinematic viscosity with respect to the Alfvén speed. If the effect of the kinematic viscosity is dominant, the reflecting layer acts like an absorbing and reflecting layer. Also, if the Alfvén speed is dominant, the reflecting layer acts like a totally reflecting barrier because the magnetic field does not produce dissipation mechanisms. Consequently, for large $|\epsilon|$, the effect of thermal conduction can be eliminated, and equation (11) will reduce to:

$$\left((D^2 - D + \frac{\sigma^2}{4}) + \alpha e^z (D^2 + D)\right) T(z) = 0. \tag{1}$$

For the solution to this equation, see [3, 30].

5. In sections 4 and 5, we introduced two solutions to describe the wave evolution, reflection, adn dissipation. The first solution was found by Laplace Transform, while the second was found by matching the inner and and outer solutions in an overlapping domain. The solution by Laplace Transform describes the behavior of the solution, which satisfies the prescribed boundary conditions, without given the same conclusions because we have only one reflecting layer and two regions. The second solution, however, gives more information, two reflecting layers, three distinct regions, the nature of every reflecting layer, and it describes the evolution in each region. Thus, we can say that the behavior and existence of the solution, and choice of the boundary conditions are very essential in the formulation of the problem, and to the existence of the solution.

References

[1] Alkahby, H. and Yanowitch. M., The effect of Newtonian cooling on the refection of vertically propagating acoustic waves in an isothermal atmosphere, *Wave Motion* 11, 419-426 (1989).

[2] Alkahby, H., Yanowitch, M., Reflection of vertically propagating waves in a thermally conducting isothermal atmosphere with a horizontal magnetic field, *Geophys. Astroph. Fluid. Dynam.* Vol. 56, (1991), 227 - 235.

[3] Alkahby, H., On the coronal heating mechanism by the resonant absorption of Alfven waves, Internat. *J. Math. and Math. Sci.*, Vol 16, No. 4, (1993), 811-816.

[4] Alkahby, H., Hussain, T., On the heating of the solar corona by the resonant absorption of Alfven waves, *Appl. Math. Lett.*, Vol 6, No. 6, (1993), 59 - 63.

[5] Alkahby, H., Reflection and dissipation of hydro-magnetic waves in a viscous and thermally conduction isothermal atmosphere, *Geophys. Astrphys. Fluid Dynam.* 72, 197-207 (1993).

[6] Alkahby, H. Y., Acoustic-gravity waves in a viscous and thermally conduction isothermal atmosphere(Part I for large Prandlt number). Internat. *J. Math Math Sci.* Vol. 18, No. 3 (1995a) 371-383.

[7] Alkahby, H. Y. M. Mahrous, and L. Debnath, The effect of Newtonian cooling on the reflection and dissipation of hydrom-agnetic waves in a viscous and thermally conducting isothermal atmosphere. *Geophys. Astroph. FLuid Dynam.* Vol. 81 (1995c) 57 - 57.

[8] Alkahby, H. Y. and M. Mahrous, On the reflection of Alfven waves in an ideal atmosphere. *Intern J. Math and Math Sci.* Vol. 21. No. 2 (1998) 381-386.

[9] Alkahby,H. Y., Talmadge, A., and Abraham, F. J. Effect of radiative damping on Magneto-hydrodynamic waves in an isothermal atmosphere. *Intern J. Math and Math Sci.* Vol. 29 (2002b) 341-348.

[10] Busse, F. H., "Thermal instabilities in rapidly rotating systems," *J. Fluid Mech.* Vol. 44, (1970) 441-460.

[11] Busse, F. H. and Carrigan, C.R., "Convection induced by centrifugal buoyancy." *J. Fluid Mech.* Vol 62. (1974) 579 -592.

[12] Busse, F. H. and Wicht. J., "A simple Dynamo caused by conductivity variations", *Geophys. Astrophys. Fluid Dnam.* Vol 64. (1992) 135-144.

[13] Cally, P. S., "Magneto-hydrodynamic critical levels and radiative damping", *Astron. Astrophys.* Vol 136, 1984 (121-126).

[14] Cally, P. S., "Magneto-hydrodynamic tube waves: waves in fibrils", *Aust. J. Phys.* Vol 38, 1985 (825 - 827).

[15] Cally, P. S., "Leaky and non-leaky oscillations in magnetic flux tubes", *Solar Phys.* Vol. 103, 1986 (121-126).

[16] Campos, L. M. B. C., "On magneto acoustic-gravity waves propagating or standing vertically in an atmosphere", *J. Phys.* Vol 59, 1987 (417-437).

[17] Ford, W. B., The asymptotic development of functions defined by Maclaurin's series, New York, Chelsea Publishing Company, 1960.

[18] Liao, X., Zhang, K., Chang, Y., Convection in rotating annular channels heated from below: Part 1. Linear stability and weakly nonlinearly mean flows, *Geophys. Astrophys. Fluid Dynamics*, Vol 99., 2006, 445-465.

[19] Lyons, P., Yanowitch, M., Vertical oscillations in a viscous and thermally conducting isothermal atmosphere., *J. Fluid Mech.*, Vol. 66, 1974, 273-288.

[20] Lyubimov, D.V., Busse, F.H., Rotating magnetic field effect on an onset convection in a horizontal layer of conducting fluid. *European Journal of Mechanics B/Fluids*. Vol 61, 2017, 16-24.

[21] Parker, E. N., Cosmical Magnetic Fields, their origin and their activity, Clarendon Press, Oxford (1979).

[22] Parker, E. N., "Sunspots and the physics of magnetic flux tubes VIII: Overstability in a magnetic field in a downdraft", *Astrophys. J.* Vol 233, 1979, 1005-1015.

[23] Parker, E. N., "Heating of solar corona holes", *Astrophys. J.* Vol 372, 1991, 719-727.

[24] Priest, E. R., Solar Active Regions (Ed. F. Orralll). Vol 213, *Colorado Assoc. Univ. Press* (1981).

[25] Priest, E. E., Solar Magnetohydrodynamics. *D. Reeidel Pub. Co.*, (1984).

[26] Simitev, R. D., Busse, F.H., Baroclinically-driven flows and dynamo actions in rotating spherical fluid shells., *Geophys. & Astrophys. Fluid Dynam.* Vol. 111., 2017, 369 - 379.

[27] Soward, A. M., "Non-linear marginal convection in a rotating magnetic system", *Geophys. Astrophys. Fluid Dynam.* Vol 35, 1986, 329-371.

[28] Yanowitch, M., "Effect of viscosity on vertical oscillations of an isothermal atmosphere". *Can. J. Phys.* Vol 45, 1967b, 2003-2008.

[29] Yanowitch, M., "Effect of viscosity on gravity waves and the upper boundary condition", *J. Fluid Mech.* Vol 29, 1967b, 209 - 231.

[30] Yanowitch, M., "A numerical study of vertically propagating waves in viscous isothermal atmosphere", *J. Comp. Phys., Vol. 4, (1969), 531 - 542.*

[31] Yanowitch, M., "Vertically propagating hydro-magnetic waves in an isothermal atmosphere with a horizontal magnetic field", *Wave Motion* Vol 1, 1979, 123-125.

In: Evolutionary Processes and Applications ISBN: 978-1-53616-350-6
Editor: G. M. N'Guérékata

Chapter 3

ERGODIC-TYPE THEOREM FOR QUANTUM DYNAMICAL SEMIGROUP

Michael O. Ogundiran*
Department of Mathematics,
Obafemi Awolowo University, Ile-Ife, Nigeria

Abstract

In this work, we shall be concerned with quantum dynamical semigroup on von Neumann algebra. The existence result for a long-time asymptotic behaviour of quantum dynamical semigroup of certain open quantum systems was established. We first established the existence of solution for Cauchy problem involving an infinitesimal generator of such semigroup then an ergodic-type theorem for the quantum dynamical semigroup was established.

Keywords: semigroup of operators, Cesaro means, completely positive maps, quantum dynamical semigroup

AMS Subject Classification: 46A35, 81S22

1. INTRODUCTION

In the description of non conservative dynamics on quantum systems, completely positive maps and their semigroups play a significant role. It is natural to assume that the master equation governing the reduced dynamics of

*Corresponding Author's Emails: mogundiran@oauife.edu.ng; adeolu74113@yahoo.com.

a physical subsystem interacting with a reservoir is the differential form of a contractive semigroup of completely positive maps. Quantum dynamical semigroup of completely positive maps had been studied extensively by authors [1], [2], [3] and the references cited in them. One of the motivations for studying quantum stochastic differential equation is its application in quantum open systems. This is achieved by the quantum markov semigroups described by the unitary solution of such differential equation, [4], [5]. The mild solution of quantum stochastic evolutions arising from quantum stochastic differential inclusions was established in [6]. A detailed work on such quantum stochastic calculus is in the monograph [7].

In [8], the quantum markov semigroup and their stationary states were studied. The convergence of the cesaro means of quantum dynamical semigroup was established. The aim of this work is to establish the existence result for a Cauchy problem involving the infinitesimal generator of quantum dynamical semigroup and an ergodic-type theorem for the semigroup which is an extension of similar results in the literatures.

2. Preliminaries

In this section we shall give the definitions and state some propositions that shall be explored in the sequel.

Definition 2.1. *Let $\mathcal{H}$ be a Hilbert space and $\mathcal{A} = B(\mathcal{H})$ be the von Neumann algebra of all bounded linear operators on $\mathcal{H}$.*
(a) A Quantum Dynamical Semigroup (QDS) is a w^*-continuous semigroup $T = \{T_t\}_{t \geq 0}$ of normal completely positive linear maps on $\mathcal{A}$ satisfying that $T_0(.)$ is the identity map of $\mathcal{A}$ and such that $T_t(1) = 1$ for all $t \geq 0$, where 1 denotes the unit operator in $\mathcal{A}$.
(b) The infinitesimal generator of T is the operator A with domain $D(A)$ which is a vector space satisfying the property that for each $a \in \mathcal{A}$,

$$w^* - \lim \frac{1}{t}\big(T_t(a) - a\big) \quad \textit{exists.}$$

(c)Let $\mathcal{A}$ be a von-Neumann algebra, and $\{a_n\}_n$ a bounded sequence of elements in $\mathcal{A}$, we define the function ϕ on $\mathcal{A}$ by

$$\phi(b) = \lim_{n\to\infty} \sup \parallel a_n - b \parallel^2 .$$

The unique point $a_\infty \in \mathcal{A}$ that minimizes ϕ on $\mathcal{A}$ is called the asymptotic centre of the bounded sequence of elements $\{a_n\}_n$.

Proposition 2.1. *The asymptotic centre of a bounded sequence of elements $a_n \in \mathcal{A}$ belongs to the closed convex hull of its weak cluster points. Moreover, if the bounded sequence of elements a_n converges weakly, then the limit coincides with the asymptotic centre*

Proof. Let a_∞ be the asymptotic centre of the bounded sequence of elements a_n and b_∞ be the projection of a_∞ onto the closed convex hull C of the weak cluster points of the sequence $\{a_n\}_n$. It is non-empty because any bounded sequence is weakly relatively compact. There exists a subsequence of elements a_n, such that $\phi(b_\infty) = \lim_{n' \to \infty} \| a_{n'} - b_\infty \|^2$. Hence

$$\limsup_{n' \to \infty} \| a_{n'} - a_\infty \|^2 \leq \phi(x_\infty).$$

Also there exists a subsequence of elements $a_{n''}$ that converges to some $z \in C$. Therefore

$$\limsup_{n'' \to \infty} \langle b_\infty - a_\infty, b_\infty - a_{n''} \rangle = \langle b_\infty - a_\infty, b_\infty - z \rangle \leq 0$$

because b_∞ is the projection of a_∞ to C.
From the identity

$$\| a_n - a_\infty \|^2 = \| a_n - b_\infty \|^2 + \| b_\infty - a_\infty \|^2 + 2\langle a_n - b_\infty, b_\infty - a_\infty \rangle$$

we deduce that

$$\phi(a_\infty) \geq \phi(b_\infty) + \| a_\infty - b_\infty \|^2 \geq \phi(b_\infty).$$

From the uniqueness of the minimum of ϕ, it follows that $a_\infty = b_\infty \in C$.
The second part of the proposition is an immediate consequence of the first.

Definition 2.2. *(1) A point x_* is a stationary point of a semigroup of operators $\{T_t\}$ if and only if*

$$T_t x_* = x_* \qquad \forall t \geq 0$$

(2) Let K be a closed convex subset of x_* and $D \subseteq \mathbb{R}_+$, a family of maps $\{T_t\}$ from K to K is called a semigroup of non-expansive maps if
(i) $T_0 = I, \qquad \forall t, s \in D \qquad T_{t+s} = T_t T_s.$

(ii) $\forall t \in D,\ \ x, y \in K, \| T_t x - T_t y \| \leq \| x - y \|$.
If x_* is a stationary point of the semigroup, the function $t \rightarrow \| T_t x - x_* \|$ is non-increasing. When $t \geq s$,

$$\| T_t x - x_* \| = \| T_{t-s} T_s x - T_{t-s} x_* \| \leq \| T_s x - x_* \| .$$

Therefore if there exists at least one stationary point x_*, the function $t \rightarrow \| T_t x - x_* \|^2$ has a limit as $t \rightarrow \infty$; Hence $t \rightarrow T_t x$ is bounded and thus it has an asymptotic centre.

Proposition 2.2. *If all the weak cluster points of $T_t x$ are stationary points, then $T_t x$ converges weakly to its asymptotic centre as $t \rightarrow \infty$.*

Proof. $M = \{y \in X : w^* - \lim_{t \rightarrow \infty} \| T_t x - y \| \ exists\}$ and N denotes the set of stationary points x_* of the semigroup; then

$$\| T_t x - x_* \| = \| T_{t-s} T_s x - T_{t-s} x_* \| \leq \| T_s x - x_* \| \quad t \geq s;$$

implies that $N \subset M$. The result therefore follows from proposition (2.1).
The following proposition gives a sufficient condition for cluster points to be stationary points.

Proposition 2.3. *Let T be a non-expansive map from a closed convex subset $K \subset \mathcal{A}$ to itself. Assume that a sequence of elements $x_N \in K$ satisfies the condition*

$$\lim_{N \rightarrow \infty} \| x_N - T x_N \| = 0. \tag{2.1}$$

Then any weak cluster point of this sequence is a fixed point of T.

Proof. Let x_* be the weak limit of a generalized subsequence $\{x_{N'}\}_{N'}$ of the sequence $\{x_N\}_N$ and let us set $x_\lambda = (1 - \lambda) x_* + \lambda T x_*$ where $\lambda \in (0, 1]$. Since T is non-expansive, we deduce that

$$\begin{aligned} \langle x_\lambda - T x_\lambda - (x_{N'} - T x_{N'}), x_\lambda - x_{N'} \rangle &\geq \| x_\lambda - x_{N'} \|^2 - \| T x_\lambda - T x_{N'} \| \| x_\lambda - x_{N'} \| \\ &\geq 0 \end{aligned}$$

By letting $x_{N'} \rightarrow x_*$, we deduce from the inequality

$$\langle x_\lambda - T x_\lambda, x_\lambda - x_* \rangle \geq 0$$

that $x_{N'} \rightarrow T x_{N'}$ strongly.
This latter inequality can be written, after division by $\lambda > 0$,

$$\langle (1 - \lambda) x_* + \lambda T x_* - T[(1 - \lambda) x_* + \lambda T x_*], T x_* - x_* \rangle \geq 0.$$

By letting $\lambda \to 0$, we deduce that $\| Tx_* - x_* \|^2 \leq 0$, which implies that $x_* = Tx_*$

Definition 2.3. *Let* $D = [0, +\infty)$*, consider a function* $Q : D \times D \to D$ *satisfying the properties*

$$\begin{aligned} &(i)\forall N \in D, \quad s \longmapsto Q(N,s) \;\; is \;\; \mu - measurable \\ &(ii)\forall N \in D, \quad \int_D Q(N,s)d\mu(s) = 1. \end{aligned} \tag{2.2}$$

If we set

$$Q(N,t) = \begin{cases} \frac{1}{N}; & when\, 0 \leq t \leq N, \\ 0; & when\, t > N. \end{cases}$$

Assuming $t \longrightarrow T_t x$ *is strongly* μ*-measurable, we define the Cesaro means* $\sigma_Q(N)x$ *by*

$$\sigma_Q(N)x = \int_D Q(N,s)T_s x d\mu(s).$$

3. Existence Results

Lemma 3.1. *Let* $\{T_t; t \geq 0\}$ *be a strongly continuous quantum dynamical semigroup of operators on the von Neumann algebra* $\mathcal{A} = B(\mathcal{H})$ *and* A *is its infinitesimal generator with domain* $D(A)$*. If* $x \in D(A)$*, then* $T_t x \in D(A)$*,* $0 \leq t < \infty$ *is strongly differentiable in t and*

$$\begin{aligned} \left(\frac{d}{dt}\right)T_t x &= AT_t x \\ &= T_t Ax, \quad t \geq 0 \end{aligned} \tag{3.1}$$

Proof. Since $T_t T_h(.) = T_h T_t(.)$, we have

$$\begin{aligned} A_h T_t x &= T_t A_h x \\ &\to T_t Ax, \;\; \text{as } h \to 0^+. \end{aligned}$$

hence $T_t x \in D(A)$ and $AT_t x = T_t Ax$ where

$$\begin{aligned} A_h x &= \frac{(T_h x - x)}{h}, \;\; x \in X \;\; \text{and} \\ Ax &= \lim_{h \to 0^+} A_h x, \;\; x \in D(A). \end{aligned}$$

Next we will show that $(\frac{d}{dt})T_t x = T_t Ax$ by showing that this is true both for the left and right hand derivatives. Indeed,

$$\lim_{h\to 0^+}\left(\frac{[T_{t+h}x - T_t x]}{h} - T_t Ax\right) = \lim_{h\to 0^+}\left[T_t A_h x - T_t Ax\right] = 0$$

and

$$\begin{aligned}
&\lim_{h\to 0^-}\left(\frac{[T_{t+h}x - T_t x]}{h} - T_t Ax\right)\\
&\quad = \lim_{h\to 0_-}\left[T_{t+h}\left\{\frac{[T_{-h}x - x]}{-h} - Ax\right\} + [T_{t+h} - T_t]Ax\right] = 0
\end{aligned}$$

hence the lemma.

Theorem 3.1. *Let A on $D(A)$ be the infinitesimal generator of a strongly continuous semigroup $\{T_t(.) : t > 0\}$. Then the Cauchy problem*

$$\begin{aligned}
\frac{dx}{dt} &= Ax, \quad t > 0\\
x(0) &= x_0, \quad x_0 \in D(A)
\end{aligned}$$

has a unique solution $x(t) = T_t x_0, \quad t > 0$

Proof. The existence of the solution follows from Lemma (3.1).

$$\begin{aligned}
(\frac{d}{dt})x(t) &= (\frac{d}{dt})T_t x_0\\
&= AT_t x_0\\
&= Ax(t)
\end{aligned}$$

and $x(0) = T_0 x_0 = x_0$.
To prove the uniqueness, let $y(t)$ be any solution of the system, set $F(s) = T_{t-s}y(s)$. Since $y(s) \in D(A)$, it follows that

$$\begin{aligned}
(\frac{d}{ds})F(s) &= -AT_{t-s}y(s) + T_{t-s}y(s)\\
&= -\,AT_{t-s}y(s) + T_{t-s}Ay(s) = 0, \quad 0 \le s \le t.
\end{aligned}$$

Hence $F(s) = c$, constant for $0 \le s \le t$. In particular, $F(0) = F(t)$, since $F(0) = T_t y(0) = T_t x_0 = x(t)$ and $F(t) = T_0 y(t) = y(t)$, which implies

$x(t) = y(t)$, and hence the proof.

Ergodic Theorem for Quantum Dynamical Semigroup (QDS)

We next establish the weak convergence of Cesaro means $\frac{1}{N}\int_0^N T_t x dt$. The following Lemmas follow from Propositions (2.1) and (2.2).

Lemma 3.2. *Assume that there exists at least one stationary point of the QDS* $\{T_t\}$ *and that*

$$\forall t \geq 0, \quad \lim_{N\to\infty} \| \sigma_Q(N)x - T_t\sigma_Q(N)x \| = 0 \tag{3.2}$$

when $N \to \infty$, $\sigma_Q(N)x$ *converges weakly to the asymptotic centre of* $t \to T_t x$, *which is a stationary point of the QDS .*

Proof. Set $y_N = \sigma_Q(N)x$ Proposition (2.2) implies that all the weak cluster points of the sequence $\{y_N\}$ are stationary points of the semigroup; these weak cluster points belong to the subset

$$K = \{y \in \mathcal{A} : \lim_{t\to\infty} \| T_t x - y \| \quad \text{exists}\}.$$

Proposition (2.1) implies that all the weak cluster points belong to the closed convex hull C of the weak cluster points of $\{x_N\}$. So, the weak cluster points of $\{y_N\}_N$ are contained in $N \cap C$ and our lemma ensues by proposition (2.2). □.

We next prove condition (3.2).

Lemma 3.3. *Set* $y_N = \sigma_Q(N)x$. *Then*

$$\begin{aligned} \| T_t y_N - y_N \|^2 \leq & \int_{D\cap[0,t]} Q(N,s) \| T_s x - T_t y_N \|^2 d\mu(s) \\ & + \int_D | Q(N,s+t) - Q(N,s) | \| T_s x - y_N \|^2 d\mu(s) \end{aligned} \tag{3.3}$$

Proof. Since $\| T_{t+s}x - T_t y \| \leq \| T_s x - y \|$, we deduce that

$$\begin{aligned} 0 \leq & \| T_s x - T_t y + T_t y - y \|^2 - \| T_{t+s}x - T_t y \|^2 \\ & + \| T_s x - T_t y \|^2 + \| T_t y - y \|^2 - \| T_{t+s}x - T_t y \|^2 \\ & + 2\langle T_s x - T_t y, T_t y - y\rangle. \end{aligned}$$

We now take $y = y_N$. The above inequality becomes

$$\| T_t y_N - y_N \|^2 \leq \int_D Q(N,s)(\| T_s x - T_t y_N \|^2 - \| T_{t+s} x - T_t y_N \|^2) d\mu(s).$$

Since

$$\int_D Q(N,s) \| T_s x - T_t y_N \|^2 \, d\mu(s) = \int_{D\cap[0,t]} Q(N,s) \| T_s x - T_t y_N \|^2 \, d\mu(s) + \int_D Q(N, s+t) \| T_{t+s} x - T_t y_N \|^2 \, d\mu(s).$$

The above inequality can be rewritten as

$$\| T_t y_N - y_N \|^2 \leq \int_{D\cap[0,t]} \| T_s x - T_t y_N \|^2 \, d\mu(s) + \int_D | Q(N,s) - Q(N,s+t) | \| T_{t+s} x - T_t y_N \|^2 \, d\mu(s)$$

We estimate the last integral by using

$$\| T_{t+s} x - T_t y \|^2 \leq \| T_s x - y_N \|^2 .$$

Hence inequality (3.3) is established. □

Theorem 3.2. *Let $T_t : K \to K$ be μ-measurable quantum dynamical semigroup of nonexpansive maps which has at least one stationary point. Let Q be any function from $D \times D$ to $\mathbb{R}_+$, measurable with respect to the second variable, of bounded variation, satisfying*

$$\begin{aligned} &(i) \forall\, N > 0, \int_D Q(N,s) d\mu(s) = 1, \\ &(ii) \forall\, t \in D, \lim_{n\to\infty} \int_{D\cap[0,t]} Q(N,s) d\mu(s) = 0, \\ &(iii) \forall\, t \in D, \lim_{N\to\infty} \int_D | Q(N,s+t) - Q(N,s) | \, d\mu(s) = 0 \end{aligned} \tag{3.4}$$

Then the average $\sigma_Q x$ converge weakly to the asymptotic centre of $t \to T_t x$ which is a stationary point of the semigroup, as $N \to \infty$.

Proof. By proposition (2.3), we have to check that

$$\forall t \geq 0 \quad \| T_t y_N - y_N \|^2 = 0$$

We use Lemma (3.3) to prove this. Observe that if $x_* \in K$ is a stationary point of the semigroup, we can estimate $\| T_s y - z \| = \| T_s y - T_s x_* + x_* - z \|$ by

$$\| T_s y - T_s x_* \| + \| x_* - z \| \leq \| x_* - y \| + \| x_* - z \| .$$

We also observe that

$$\| x_* - y_N \| = \| \int_D Q(N,s)(T_s x_* - T_s x) d\mu(s) \| \leq \| x_* - x \| .$$

By taking $y = x$ and $z = T_t y_N$, we deduce that

$$\begin{aligned} \| T_s x - T_t y_N \| &\leq \| x_* - x \| + \| x_* - T_t y_N \| \\ &\leq \| x_* - x \| + \| x_* - y_N \| \\ &\leq 2 \| x_* - x \| \end{aligned}$$

and by taking $y = x$ and $z = y_N$, we deduce that

$$\| T_s x - y_N \| \leq 2 \| x_* - x \| .$$

Hence Lemma (3.3) implies that

$$\begin{aligned} \| T_t y_N - y_N \|^2 \leq 4 \| x_* - x \|^2 \Bigg(&\int_{D \cap [0,t]} Q(N,s) d\mu(s) \\ &+ \int_D | Q(N, t+s) - Q(N,s) | d\mu(s) \Bigg) \end{aligned}$$

The right hand side of the inequality converges to as $N \to \infty$ by assumption (3.4) (i) and (ii).

Corollary 3.1. *Let $T_t : K \to K$ be a quantum dynamical semigroup of non-expansive maps which has at least one stationary point. If the semigroup $T_.$ is measurable, then*

$$\forall x \in K, \frac{1}{N} \int_0^N T_\tau x d\tau$$

converges weakly to a stationary point.

REFERENCES

[1] Accardi L. and Kozyrev V., On the structure of quantum Markov flows, *Preprint Centro V. Volterra* 1999 .

[2] Chebotarev A. M. and Fagnola F., Sufficient conditions for conservativity of minimal quantum dynamical semigroups *J. Funct. Anal.* 153, (1998) , 382-404.

[3] Davies E. B., Generators of dynamical semigroups *J. Funct. Anal.*, 34, (1979), 421-432.

[4] Blanchard P. and Olkiewicz R., Decoherence as irreversible dynamical process in Open quantum systems *Open quantum systems, Recent developments* Vol. III, Springer-Verlag Berlin 2006, 117-159.

[5] Fagnola F., Quantum Markov semigroups and quantum flows, *Proyecciones, J. Math.*, 18, (3) (1999), 1-144.

[6] Ogundiran M. O., On the mild solutions of Quantum stochastic evolution inclusions, Communications in Appl. Anal., 19 (2015), 2, 307-318.

[7] Parthasarathy K. R., An introduction to Quantum stochastic calculus, *Monographs in Mathematics* Vol.85, Birkhauser-Verlag, Basel-Boston-Berlin, 1992.

[8] Fagnola F. and Rebolledo R., Notes on the Qualitative Behaviour of Quantum Markov Semigroups *Open quantum systems, Recent developments* Vol. III, Springer-Verlag Berlin 2006 , 161-205.

[9] Alicki R. and Lendi K., *Quantum Dynamical Semigropus and Applications*, Lect. Notes Phys. 286, Springer- Verlag, 1987.

[10] Bratelli O. and Robinson D. W., *Operator Algebras and Quantum Statistical Mechanics*, Vol. 1, Springer- Verlag, 2nd edition, 1987.

[11] Fagnola F., Rebolledo R. and Saavedra C., Quantum flows asssociated to master equation in quantum optics, *J. Math. Phys.*, 35, (1994), no 1, 1-12.

[12] Frigerio A. and Verri M., Long-time asymptotic properties of dynamical semigroups on w^*-algebras, *Math. Zeitschrift*,180, no. 2, (1982), 275-286.

[13] Holevo A. S., On the structure of covariant dynamical semigroups, *J. Funct. Anal.* 131 (2), (1995), 255-278.

[14] Reed M. and Simon B., *Methods of Modern Mathematical Physics II, Fourier Analysis, Self-Adjointness*, Academic Press, 1975.

In: Evolutionary Processes and Applications
Editor: G. M. N'Guérékata
ISBN: 978-1-53616-350-6

Chapter 4

ON THE STABILITY OF HIGHER ORDER DIFFERENTIAL EQUATIONS

B. S. Kalitine*
Faculty of Economics,
Belarusian State University, Minsk, Belarus

Abstract

In this paper, we consider the problem of stability of the equilibrium of a non-linear differential equation of order n. The investigations are carried out using the method of semi-definite Lyapunov's functions for the problems of stability, asymptotic stability (local and global), and instability. The method of constructing the semi-definite functions, suitable for solving problems of stability, is implicitly stated. The results of asymptotic stability of the equilibrium coincide with necessary and sufficient conditions in the corresponding linear case. Consequently, they meet generally accepted requirements. The conducted studies show that the use of semi-defined positive functions can give advantages in comparison with the classical method of application of Lyapunov's definite positive functions.

Keywords: differential equation, equilibrium, stability, semi-definite Lyapunov's function

*Corresponding Author's Email: kalitine@yandex.by.

1. INTRODUCTION

Let $\mathbb{R}^n$ denote the real Euclidian space of dimension n with the scalar product $\langle\cdot,\cdot\rangle$ and the norm $||\cdot|| = \sqrt{\langle\cdot,\cdot\rangle}$. Let $\varepsilon > 0$ denote by $B_\varepsilon = \{x \in \mathbb{R}^n : ||x|| < \varepsilon\}$ and $\mathbb{R}^+ = [0, +\infty)$. If $U \subset \mathbb{R}^n$ is an open subset, the by $C^1(U, \mathbb{R}^n)$ we denote the set of all continuously differentiable functions.

Consider the differential equations

$$\dot{x} = f(x),\ x \in G \subset \mathbb{R}^n,\ f(0) = 0, \tag{1.1}$$

where G is a neighborhood of the origin. Assume that the function $f : G \to \mathbb{R}^n$ is continuous and satisfies the conditions that ensure the uniqueness of solutions in G. The solution of (1.1) passing through the point $x_0 \in G$ at the initial moment $t = 0$ will be denoted by $x(x_0, t)$, i.e., $x(x_0, 0) = x_0$. Equation (1.1) has a trivial solution $x = 0$.

A set $Y \subset \mathbb{R}^n$ is called positively invariant whenever $x(x_0, t) \in Y$ for all $x_0 \in Y$ and $t > 0$.

We shall use the following definitions [1–8]. Let Y be a closed connected positively invariant set in $\mathbb{R}^n$ containing the origin.

The solution $x = 0$ to system (1.1) is

- Y-stable if

$$\forall\varepsilon > 0,\ \exists\delta = \delta(\varepsilon) > 0,\ \forall x_0 \in B_\delta \cap Y \Rightarrow ||x(x_0, t)|| < \varepsilon\ \forall t \geq 0;$$

- Y-attracting if

$$\exists\sigma > 0,\ \forall\alpha > 0,\ \exists T > 0,\ \forall x_0 \in B_\sigma \cap Y \Rightarrow ||x(x_0, t)|| < \alpha\ \forall t \geq T;$$

- Y-asymptotically stable if it is Y-stable and Y-attracting;

- Y-globally asymptotically stable if it is Y-asymptotically stable, $G = \mathbb{R}^n$, and $x(x_0, t) \to 0$ as $t \to +\infty$ for each $x_0 \in Y$;

- instable if

$$\exists\varepsilon > 0,\ \forall\delta > 0,\ \exists x_0 \in B_\delta,\ \exists t^* > 0 \Rightarrow ||x(x_0, t^*)|| \geq \varepsilon.$$

If $Y = \mathbb{R}^n$, then we get the usual properties of stability, attraction and asymptotic stability [9].

Let U is an open subset of G and $\bar{U}$ is the closure of U. Let $\mathbb{R}^+$ be the set of nonnegative real numbers. If $V \in C^1(U, \mathbb{R}^+)$ then we put

$$\dot{V}(x_0) = \frac{dV(x_0)}{dt} = \left\langle \frac{\partial V(x_0)}{\partial x}, f(x_0) \right\rangle$$

the time derivative of function V by virtue of the system (1.1) at the point x_0.

Now we present theorems of a direct method using the semi-definite Lyapunov's functions.

Theorem 1. [4, 5] *If there exists a neighborhood U of the origin $0 \in \mathbb{R}^n$ and a function $V \in C^1(U, \mathbb{R}^+)$ such that*

1. *$V(x) \geq 0$ for all $x \in U$ and $V(0) = 0$;*
2. *$\dot{V}(x) \leq 0$ for all $x \in U$;*
3. *solution $x = 0$ is Y_0-asymptotically stable, where $Y_0 = \{x \in \bar{U}\colon\ V(x) = 0\}$,*

then solution $x = 0$ of the system (1.1) is stable.

Theorem 2. [3, 5, 8] *If there exists a neighborhood U of the origin $0 \in \mathbb{R}^n$ and a function $V \in C^1(U, \mathbb{R}^+)$ such that*

1. *$V(x) \geq 0$ for all $x \in U$ and $V(0) = 0$;*
2. *$\dot{V}(x) \leq 0$ for all $x \in U$;*
3. *solution $x = 0$ is Y-asymptotically stable, where $Y = \{x \in \bar{U}\colon \dot{V}(x) = 0\}$,*

then solution $x = 0$ of the system (1.1) is asymptotically stable.

Theorem 3. [3, 5, 8] *Let $G = \mathbb{R}^n$. If there exists a function $V \in C^1(\mathbb{R}^n, \mathbb{R}^+)$ such that*

1. *$V(x) \geq 0$ for all $x \in \mathbb{R}^n$ and $V(0) = 0$;*
2. *$\dot{V}(x) \leq 0$ for all $x \in \mathbb{R}^n$;*
3. *solution $x = 0$ is Y_∞-asymptotically stable, where $Y_\infty = \{x \in \mathbb{R}^n : \dot{V}(x) = 0\}$,*
4. *all solutions of the system (1.1) are bounded for $t > 0$,*

then solution $x = 0$ of the system (1.1) is globally asymptotically stable.

Remark 4. *It's obvious that the set Y_0 in Theorems 1.1 – 1.3 is positively invariant. Consequently, in the specific case of using the Lyapunov sign-positive function with a sign-negative derivative with respect to time, the instability of equilibrium relative to the set Y_0 implies its instability with respect to the whole phase space.*

2. The Differential Equation of the nth Order

In this section, we study the problem of stability of a trivial solution for some class of higher order differential equations using so-called semi-definite functions.

Let $n \in \mathbb{N}$ be a natural number and $x(t)$ $(x : \mathbb{R} \to \mathbb{R})$ is n times continuously differentiable function. We shall use the following notation for derivatives of higher orders

$$\dot{x} = \frac{dx}{dt},\ \ddot{x} = \frac{d^2x}{dt^2}, \ldots,\ x^{(k)} = \frac{dx^{(k-1)}}{dt},\ k = \overline{1,n}\ (x^{(0)} = x).$$

Consider a nonlinear scalar n^{th} order differential equation

$$x^{(n)} + \sum_{i=1}^{n-1} f_i(x, \dot{x}, \ddot{x}, \ldots, x^{(n-1)})x^{(n-i)} + f_n(x, \dot{x}, \ddot{x}, \ldots, x^{(n-1)})x = 0,\ x \in \mathbb{R}, \tag{2.1}$$

where the functions $f_i : \mathbb{R}^n \to \mathbb{R},\ i = \overline{1,n}$, are continuous and satisfies the conditions that ensure the uniqueness of solutions. We put $X = (x, \dot{x}, \ddot{x}, \ldots, x^{(n-1)})$ and suppose that there exist numbers $\alpha_i \in \mathbb{R},\ i = \overline{1,n}$, for which the following identities hold

$$\begin{cases} f_i(X) - \alpha_i = \alpha_{i-1}(f_1(X) - \alpha_1),\ i = \overline{2, n-1}, \\ f_n(X) = \alpha_{n-1}(f_1(X) - \alpha_1). \end{cases} \tag{2.2}$$

We study the problem of stability of equation (2.1). To do this, we make the change of variables

$$\begin{aligned} &\dot{x} = x_1,\ \ddot{x} = x_2,\ \ldots,\ x^{(n-2)} = x_{n-2}, \\ &x^{(n-1)} + \sum_{i=1}^{n-1} \alpha_i x^{(n-1-i)} = x_{n-1}. \end{aligned} \tag{2.3}$$

The first group of equations gives $(n-2)$ first-order differential equations

$$\dot{x} = x_1,\ \dot{x}_1 = x_2,\ \ldots,\ \dot{x}_{n-3} = x_{n-2}.$$

Now we derive the differential equation with respect to the variable x_{n-2}. It follows from (2.3) that $x_{n-2} = x^{(n-2)}$. Differentiating both sides and using the last of equalities (2.3), we obtain

$$\dot{x}_{n-2} = x^{(n-1)} = x_{n-1} - \sum_{i=1}^{n-1} \alpha_i x^{(n-1-i)} = x_{n-1} - \sum_{i=1}^{n-2} \alpha_i x^{(n-1-i)} - \alpha_{n-1} x. \tag{2.4}$$

Using the expressions for the derivatives $x^{(n-1-i)}$ of (2.3) here, we can write

$$\dot{x}_{n-2} = -\alpha_{n-1} x - \sum_{i=1}^{n-2} \alpha_{n-1-i} x_i + x_{n-1}.$$

Finally, we derive the equation with respect to the variable x_{n-1}. To this end, we differentiate the last of equalities (2.3) with respect to time t. Then, taking into account the representation of the derivative $x^{(n)}$ in equation (2.1), we'll have

$$\dot{x}_{n-1} = x^{(n)} + \sum_{i=1}^{n-1} \alpha_i x^{(n-i)} = -\sum_{i=1}^{n-1} f_i x^{(n-i)} - f_n x + \sum_{i=1}^{n-1} \alpha_i x^{(n-i)},$$

$$= -(f_1 - \alpha_1) x^{(n-1)} - \sum_{i=2}^{n-1} (f_i - \alpha_i) x^{(n-i)} - f_n x.$$

We replace here the function $f_i(X)$ of the identities (2.2). Then we obtain a differential equation with respect to the variable x_{n-i} by the following transformations

$$\dot{x}_{n-1} = -(f_1 - \alpha_1) x^{(n-1)} - \sum_{i=2}^{n-1} \alpha_{i-1} (f_1 - \alpha_1) x^{(n-i)} - \alpha_{n-1} (f_1 - \alpha_1) x$$

$$= -(f_1 - \alpha_1) \left(x^{(n-1)} + \sum_{i=2}^{n-1} \alpha_{i-1} x^{(n-i)} + \alpha_{n-1} x \right)$$

$$= -(f_1 - \alpha_1) \left(x^{(n-1)} + \sum_{i=2}^{n} \alpha_{i-1} x^{(n-i)} \right)$$

$$= -(f_1 - \alpha_1) \left(x^{(n-1)} + \sum_{i=1}^{n-1} \alpha_i x^{(n-1-i)} \right) = -(f_1 - \alpha_1) x_{n-1}.$$

Thus, the change of variables (2.3) leads us to a system of differential equations

$$\begin{cases} \dot{x} = x_1, \ \dot{x}_1 = x_2, \ ..., \ \dot{x}_{n-3} = x_{n-2}, \\ \dot{x}_{n-2} = -\alpha_{n-1}x - \sum\limits_{i=1}^{n-2} \alpha_{n-i-1}x_i + x_{n-1}, \\ \dot{x}_{n-1} = -(\varphi(x, x_1, x_2, ..., x_{n-1}) - \alpha_1)x_{n-1}, \end{cases} \tag{2.5}$$

where on the basis of (2.4) we use the notation

$$\varphi(x, x_1, x_2, ..., x_{n-1}) = f_1\left(x, x_1, x_2, ..., x_{n-2}, x_{n-1} - \sum_{i=1}^{n-2} \alpha_i x_{n-i-1} - \alpha_{n-1}x\right).$$

It is clear that the problem of the stability of the solution $x = \dot{x} = \ddot{x} = ... = x^{(n-1)} = 0$ of equation (2.1) is equivalent to the problem of stability of the solution $x = x_1 = x_2 = ... = x_{n-1} = 0$ of the system (2.5). To investigate the stability problem for system (2.5), we take a simple semi-definite positive function

$$V(x, x_1, x_2, ..., x_{n-1}) = \frac{1}{2}(x_{n-1})^2.$$

Its time derivative by virtue of the system (2.5) is equal to

$$\dot{V}(x, x_1, x_2, ..., x_{n-1}) = -(\varphi(x, x_1, x_2, ..., x_{n-1}) - \alpha_1)(x_{n-1})^2.$$

In order to obtain stability conditions for the zero solution, we make two assumptions about the system (2.5).

1) Suppose that there exists a number $h > 0$ such that

$$\varphi(x, x_1, x_2, ..., x_{n-1}) \geq \alpha_1 \ for \ all \ ||(x, x_1, x_2, ..., x_{n-1})|| < h.$$

According to this condition, the derivative $\dot{V}(x, x_1, x_2, ..., x_{n-1})$ is not positive in the h-neighborhood of the origin.

2) Let on the set, where $V(x, x_1, x_2, ..., x_{n-1}) = 0$, (that is, for $x_{n-1} = 0$), the zero solution of the system (2.5) is asymptotically stable. On this set, the system takes the form

$$\begin{cases} \dot{x} = x_1, \ \dot{x} = x_2, \ ..., \ \dot{x}_{n-3} = x_{n-2}, \\ \dot{x}_{n-2} = -\alpha_{n-1}x - \sum\limits_{i=1}^{n-2} \alpha_{n-i-1}x_i. \end{cases} \tag{2.6}$$

If Assumptions 1) and 2) are satisfied, all the conditions of Theorem 1.1 for system (2.5) will be satisfied, and hence the zero solution $x = \dot{x} = \ddot{x} = ... =$

$x^{(n-1)} = 0$ of equation (2.1) will be stable. If, under assumption 1), we require the strict inequality

$$\varphi(x, x_1, x_2, ..., x_{n-1}) > \alpha_1 \; for \; all \; ||(x, x_1, x_2, ..., x_{n-1})|| < h$$

then the sets Y_0 and Y of points $(x, x_1, x_2, ..., x_{n-1})$, where, respectively, $V = 0$ and $\dot{V} = 0$ are coincide.

In this case, according to Theorem 1.2, the zero solution of (2.5) will be asymptotically stable. In other words, under the stipulated assumptions the zero solution of equation (2.1) is asymptotically stable.

It is easy to verify that the system (2.6) is equivalent to a following differential equation of the $(n-1)^{th}$ order with constant coefficients

$$x^{(n-1)} + \alpha_1 x^{(n-2)} + \alpha_2 x^{(n-3)} + ... + \alpha_{n-2}\dot{x} + \alpha_{n-1}x = 0.$$

Therefore, the zero solution of the system (2.6) will be asymptotically stable if all the roots of the characteristic equation

$$\lambda^{n-1} + \alpha_1 \lambda^{n-2} + \alpha_2 \lambda^{n-3} + ... + \alpha_{n-2}\lambda + \alpha_{n-1} = 0 \tag{2.7}$$

will have negative real parts. Because of our studies and remarks, we arrive at the following result.

Theorem 5. *Suppose that for a differential equation* (2.1) *there are constant numbers* $\alpha_i \in \mathbb{R}$ *and* $h > 0$ *such that the identities* (2.2) *are satisfied and the roots of equation* (8) *have only negative real parts. Then the solution* $x = \dot{x} = \ddot{x} = ... = x^{(n-1)} = 0$ *of* (2.1) *will be*

- stable if for all $||(x, x_1, x_2, ..., x_{n-1})|| < h$

$$f_1\left(x, x_1, x_2, ..., x_{n-2}, x_{n-1} - \sum_{i=1}^{n-2} \alpha_i x_{n-i-1} - \alpha_{n-1}x\right) \geq \alpha_1,$$

- asymptotically stable if for all $||(x, x_1, x_2, ..., x_{n-1})|| < h$

$$f_1\left(x, x_1, x_2, ..., x_{n-2}, x_{n-1} - \sum_{i=1}^{n-2} \alpha_i x_{n-i-1} - \alpha_{n-1}x\right) > \alpha_1.$$

If, however, one of the roots of equation (8) *has a positive real part, then the solution* $x = \dot{x} = \ddot{x} = ... = x^{(n-1)} = 0$ *of equation* (2.1) *is unstable.*

Example. Consider a fourth-order nonlinear differential equation

$$\ddddot{x} + a\dddot{x} + 3\alpha(a - 2\alpha)\ddot{x} + \alpha^2(3a - 8\alpha)\dot{x} + \alpha^3(a - 3\alpha)x = 0, \tag{2.8}$$

where $\alpha \in \mathbb{R}$ is a constant, and a function $a = a(x, \dot{x}, \ddot{x}, \dddot{x})$ $(a : \mathbb{R}^4 \to \mathbb{R})$ is continuous and ensures the uniqueness of the solutions. The problem of stability of certain fourth-order differential equations was considered by several authors (see [10]).

It is easy to verify that the identities (2.2) are satisfied for equation (2.10) if we put $\alpha_1 = 3\alpha, \alpha_2 = 3\alpha^2, \alpha_3 = \alpha^3$. In this case, the characteristic equation (2.7) is

$$\lambda^3 + 3\alpha\lambda^2 + 3\alpha^2\lambda + \alpha^3 = 0$$

with one single root $\lambda = -\alpha$. Therefore, in this case the solution $x = \dot{x} = \ddot{x} = 0$ of the differential equation (2.7) is asymptotically stable if $\alpha > 0$.

Thus, according to Theorem 1.1 and Theorem 1.2, the zero solution of (2.10) is stable if

$$\varphi(x, y, z, u) \geq 3\alpha > 0 \ for \ ||(x, y, z, u)|| < h \tag{2.9}$$

and it will be asymptotically stable if

$$\varphi(x, y, z, u) > 3\alpha > 0 \ for \ ||(x, y, z, u)|| < h, \tag{2.10}$$

where it is accepted $\varphi(x, y, z, u) = a(x, y, z, -\alpha^3 x - 3\alpha^2 y - 3\alpha u)$.

In this case the system (2.5) takes the form

$$\begin{cases} \dot{x} = y, \\ \dot{y} = z, \\ \dot{z} = -\alpha^3 x - 3\alpha^2 y - 3\alpha z + u, \\ \dot{u} = -(\varphi(x, y, z, u) - 3\alpha)u. \end{cases} \tag{2.11}$$

Let us show that, for the given example, the fulfillment of inequality (2.12) for all $(x, y, z, u) \in \mathbb{R}^4$ (that is, when $h = +\infty$) guarantees the global asymptotic stability of the solution $x = \dot{x} = \ddot{x} = \dddot{x} = 0$. In fact, consider a semidefinite function for the nonlinear system (2.13)

$$V(x, y, z, u) = 0,5u^2, \ \dot{V}(x, y, z, u) = -(\varphi(x, y, z, u) - 3\alpha)u^2. \tag{2.12}$$

On the basis of condition (2.12) (for $h = +\infty$), the derivative (2.14) will be no positive for all $(x, y, z, u) \in \mathbb{R}^4$. Therefore, in order to use Theorem 1.3, it is necessary to prove the boundedness of any solution $(x(t), y(t), z(t), u(t))$ of the system (2.13). We will show this.

The set Y on which the function $\dot{V}(x, y, z, u) = 0$ is defined by equality $u = 0$. This set is positively invariant, and on it the system (2.13) becomes a linear system of differential equations

$$\begin{cases} \dot{x} = y, \\ \dot{y} = z, \\ \dot{z} = -\alpha^3 x - 3\alpha^2 y - 3\alpha z. \end{cases} \tag{2.13}$$

The global asymptotic stability of the zero solution for (2.15) follows from the fulfillment of the condition $\alpha > 0$.

In addition, the presence of a semi-positive function (2.14) with a semi-negative derivative means that the component $|u(t)|$ is bounded for $t \geq 0$.

Let's consider one more function $V_1(x, y, z, u) = 0,5z^2$. Its derivative with respect to time in view of the system (2.13) is

$$\dot{V}_1(x, y, z, u) = -\alpha^3 xz - 3\alpha^2 yz - 3\alpha z^2 + zu \tag{2.14}$$

It follows that the component $|z(t)|$ of the solution $(x(t), y(t), z(t), u(t))$ of (2.13) will be bounded for $t \geq 0$ if the derivative $\dot{V}_1(x, y, z, u)$ is negative for sufficiently large $|z(t)|$. Taking into account the boundedness with respect to the component $|u(t)|$ and the condition $\alpha > 0$, the inequality $\dot{V}_1(x, y, z, u) < 0$ will hold if for sufficiently large $|z(t)|$ the inequalities $x(t)z(t) \geq 0$ and $y(t)z(t) \geq 0$ are satisfied.

Let us show that for sufficiently large $|z(t)|$ the inequality $y(t)z(t) \geq 0$ holds. In fact, according to the second equation of the system, we have the equality $\dot{y}(t) = z(t)$. And this means that for a sufficiently large $|z(t)|$ the values of $y(t)$ and $z(t)$ will have the same signs. Consequently, for sufficiently large $|z(t)|$, the inequality $y(t)z(t) \geq 0$ holds. Moreover, for a sufficiently large $|z(t)|$ the component $|y(t)|$ will also be arbitrarily large.

Now we show that for sufficiently large $|z(t)|$ the inequality $x(t)z(t) \geq 0$ holds. We first note that, according to the first equation of the system (14), namely $\dot{x}(t) = y(t)$, it follows that for a sufficiently large $|y(t)|$ the values of $y(t)$ and $x(t)$, they will have the same signs. Moreover, for a sufficiently large $|y(t)|$ the component $|x(t)|$ will also be arbitrarily large and $x(t)y(t) \geq 0$.

Consequently, for sufficiently large values $|z(t)|$ of the inequalities $x(t)y(t) \geq 0$ and $y(t)z(t) \geq 0$ it follows that $x(t)z(t) \geq 0$.

Thus, for a sufficiently large value $|z(t)|$, the derivative $\dot{V}_1(x, y, z, u) < 0$ which means the boundedness of the solution $(x(t), y(t), z(t), u(t))$ with respect to the component $|z(t)|$.

It remains to prove the boundedness with respect to the components $y(t)$ and $x(t)$. Suppose on the contrary that the component $y(t)$ is not bounded. Then, according to the previous arguments, the component $x(t)$ is also unbounded, and the inequality $x(t)y(t) \geq 0$ holds. This means that for a sufficiently large $y(t)$ the derivative (2.16) will be unbounded. Then the function $V_1(x, y, z, u) = 0,5z^2$ will increase without limit. However, this contradicts the proven boundedness of $|z(t)|$. This contradiction shows that the assumption of the unboundedness of the component $y(t)$ is false, and therefore we have the boundedness of the solution $(x(t), y(t), z(t), u(t))$ with respect to the component $y(t)$.

Finally, let us assume, by contradiction, that the solution $(x(t), y(t), z(t), u(t))$ of (2.13) is not bounded by the component $x(t)$. Then, similarly to the arguments of the preceding case, it follows that the derivative (2.16), and with it the component $|z(t)|$, is not bounded for a sufficiently large value of $x(t)$. This, as in the previous case, leads to a contradiction with the boundedness of $|z(t)|$.

Thus, we have shown that every solution $(x(t), y(t), z(t), u(t))$ of the system (2.13) is bounded for $t > 0$.

We state the obtained result in the form of the following assertion.

Theorem 6. *The solution* $x = \dot{x} = \ddot{x} = \dddot{x} = 0$ *of the differential equation* (2.10) *will be*

- *stable if for some number* $h > 0$

$$a(x, \dot{x}, \ddot{x}, \dddot{x}) \geq 3\alpha > 0 \; for \; ||(x, \dot{x}, \ddot{x})|| < h,$$

- *globally asymptotically stable if*

$$a(x, \dot{x}, \ddot{x}, \dddot{x}) > 3\alpha > 0 \; for \; all \; (x, \dot{x}, \ddot{x}) \in \mathbb{R}^4.$$

If $\alpha < 0$, *then the solution* $x = \dot{x} = \ddot{x} = \dddot{x} = 0$ *of the equation* (2.10) *is unstable.*

We note that the resulting conditions for the global asymptotic stability for the equation (2.8) coincide with necessary and sufficient conditions in the corresponding linear case. In addition, if the inequality $\alpha < 0$ is satisfied, then in the linear case we will also have instability.

REFERENCES

[1] Liapunov A. M., *Stability of Motion*. Academic Press, New York, 1966.

[2] Krasovskii N. N., *Certain problems of the theory of stability of motion.* Moskow, 1959, 211 p. (in Russian), trans, Stanford, Cal., 1963.

[3] Bulgakov N. G., Kalitine B.S., A generalization of theorems of Lyapunov's second method. I. Theory. *Vesci Acad. Nauk BSSR. Ser. Fiz.-Mat. Nauk,* 3 (1978) pp. 32 - 36. (in Russian).

[4] Kalitine B. S., On the theorem for nonasymptotic stability in the direct Lyapunov's method. *C. R. Acad. Sci., Paris, Ser. I* 338 (2004), P. 163-166.

[5] Kalitine B.S., *Stability of differential equations (method of semi definite Lyapunov functions).* LAP Lambert Academic Publishing, Saarbrücken, 2012, 230 p. (in Russian).

[6] Kalitine B. S., On the stability of positively invariant compact sets of dynamical systems. *R.A.I.R.O. Automatique/Systems Analysis and Control,* 16 (3) (1982), P. 275-286.

[7] Kalitine B.S., *Stability of non-autonomous differential equations).* Belarus State Univ., Minsk, 2013, 227 p. (in Russian).

[8] Kalitine B.S., *On problems of stability of non-autonomous differential equations (method of semi definite Lyapunov functions).* LAP Lambert Academic Publishing, Saarbrücken, 2015, 269 p. (in Russian).

[9] Rouch N., Habets P., Laloy M., *Stability Theory by Liapounov's Direct Method.* Springer, Berlin, 1977, 300 p.

[10] Barbachin E.A., *Lyapunov functions.* Moskov, 1970, 240 p. (in Russian).

In: Evolutionary Processes and Applications
Editor: G. M. N'Guérékata
ISBN: 978-1-53616-350-6

Chapter 5

Asymptotics Related to Nonlinear Effects for a Class of Higher Order Parabolic Equations

Mahmoud Qafsaoui*
Ecole Supérieure des Techniques Aéronautiques et de Construction Automobile - Groupe ISAE, Laval, France

Abstract

This chapter deals with the large time behavior of solutions to the initial value problem related to the class of nonlinear parabolic equations of type

$$\begin{cases} u_t + \Delta^m(a\Delta^m u) = \nabla^\theta.\mathcal{J}\big(\nabla^\nu u\big), \\ u(.,0) = u_0 \in L^p(\mathbb{R}^n). \end{cases} \tag{0.1}$$

The vector field $\mathcal{J}$ represents a nonlinearity and the bounded measurable nonnegative function a is independent of time.

Keywords: higher order elliptic operators, large time behavior, "generalized Burgers-type" nonlinearity

*Corresponding Author's Email: mahmoud.qafsaoui@estaca.fr.

1. Introduction and Statement of the Main Results

The main objective of this work is to investigate the solutions to the initial value problem (0.1), where $p \geq 1$, $\mu \in \mathbb{N}$, $\theta \in \mathbb{N}^*$ and ∇^k denotes the vector $\left(D^\gamma\right)_{|\gamma|=k}$ with $\gamma = (\gamma_1, \cdots, \gamma_n) \in \mathbb{N}^n$. The real function $u : (x,t) \longmapsto u(x,t)$ is defined on $\mathbb{R}^n \times (0,\infty)$ and the nonlinear $\mathcal{C}^\theta$ vector field $\mathcal{J} : \mathbb{R}^k \longmapsto \mathbb{R}^{k'}$, where $k = \frac{(\mu+n-1)!}{\mu!(n-1)!}$ and $k' = \frac{(\theta+n-1)!}{\theta!(n-1)!}$, is such that $|\mathcal{J}(X)| \leq C|X|^\omega$ for some nonnegative real ω and $\mathcal{J}(0) = 0$. We mainly deal with the *a-priori* L^q-estimates, for some rank of q, and the asymptotic behavior regarding the nonlinear effects.

Before going into detail, let us dwell on some work related to these types of equations which inspired ours. In [1], the authors considered the viscously damped conservation law

$$u_t - \Delta u = \nabla.\big(f(x,u)u(x,t)\big),$$

where f is a vector field satisfying for all $q \geq 1$, $\displaystyle\int_{\mathbb{R}^n} |\xi(x)|^q\big(\nabla.f(x,\xi)\big)dx = 0$, with ξ a smooth function of compact support. The most famous particular cases are Burger's equation and the two dimensional Navier-Stokes vorticity equation.

They obtained L^p-bounds and optimal control on the spatial and temporal decays of solutions. More precisely, they established the following optimal estimates for all $t > 0$ and all $1 \leq p \leq q \leq \infty$,

$$\|u(t)\|_q \leq C(n,p,q)t^{-n(1/p-1/q)/2}\|u_0\|_p\,,$$

with $C(n,p,q) = \Big(\big(p^{1/q}q'^{1/q'}/p'^{1/p'}q^{1/p}\big)\big((q-p)/4\pi\big)^{1/p-1/q}\Big)^{n/2}$ and where p', q' are respectively the conjugate indices of p and q.

Under the additional conditions $\nabla.f(.,u(.,t)) = 0$ and $\|f(.,u(.,t))\|_\infty \leq \beta(t)$, with β an integrable function, the solutions satisfy the following decay

$$|u(x,t)| \leq \int_{\mathbb{R}^n} G(t,x,y)|u(0,y)|dy\,,$$

where the kernel G is defined by

$$G(t,x,y) = (4\pi t)^{-n/2}\exp\left(-\frac{|x-y|^2}{4t}\max\Big(1 - \frac{1}{|x-y|}\int_0^t \beta(s)ds\,,\,0\Big)^2\right).$$

On the other hand, the temporal decay of $\|u(t)\|_1$ to 0, for $t \to \infty$, was obtained when $\displaystyle\int_{\mathbb{R}^n} u_0(x)dx = 0$ and $|f(.,u(.,t))| \leq c\|u(.,t/2)\|_1/\sqrt{t}$ for a fixed constant c.

In [2], the authors studied the large time behavior of solutions to the Cauchy problem for the Lévy conservation laws

$$u_t + \mathcal{L}u + \nabla.f(u) = 0,$$

supplemented with the initial condition $u_0 = u(.,0) \in L^1(\mathbb{R}^n)$. The function $f : \mathbb{R} \longrightarrow \mathbb{R}^n$ is a nonlinearity and the pseudodifferential operato $\mathcal{L}$ is defined via the Fourier transform by $\widehat{\mathcal{L}u}(\xi) = a(\xi)\hat{u}(\xi)$, where a is a positive-definite function. These equations, which appear in various physical applications such as phenomena with anomalous diffusions, are a generalization and an extension of the one-dimensional multifractal conservation laws

$$u_t - u_{xx} + \sum_{i=1}^{k} a_i\big(-\partial^2/\partial x^2\big)^{\frac{\alpha_i}{2}} u + f(u)_x = 0\,,$$

where $0 < \alpha_i < 2$ and $a_i > 0$. In particular, the authors showed, under $u_0 \in L^1(\mathbb{R}^n) \cap L^\infty(\mathbb{R}^n)$ and some conditions on the nonlinearity f, the existence of a unique mild solution in $\mathcal{C}\big([0,\infty); L^1(\mathbb{R}^n) \cap L^\infty(\mathbb{R}^n)\big)$, the regularity of this solution in $\mathcal{C}\big((0,\infty); W^{2,2}(\mathbb{R}^n)\big) \cap \mathcal{C}^1\big((0,\infty); L^2(\mathbb{R}^n)\big)$ and the following asymptotics satisfied for all $p \in]1,\infty]$,

$$\begin{cases} \lim\limits_{t\to\infty} \left(t^{n(1-1/p)/\alpha}\|u(t) - e^{-t\mathcal{L}}u_0\|_p\right) = 0\,, \\ \lim\limits_{t\to\infty} \left(t^{(n(1-1/p)+1)/\alpha}\|u(t) - e^{-t\mathcal{L}}u_0 + \displaystyle\int_D f(u(y,s))dyds.\nabla(e^{-t\mathcal{L}})\|_p\right) = 0 \end{cases}$$

where $D = \mathbb{R}^n \times (0,\infty)$. We recall, for the reader's convenience, that for $1 \leq p \leq \infty$ and $k \in \mathbb{N}$, the Sobolev space $W^{k,p}(\mathbb{R}^n) = \{u \in L^p(\mathbb{R}^n)/D^\gamma u \in L^p(\mathbb{R}^n)$ for all $\gamma \in \mathbb{N}^n$ such that $|\gamma| \leq k\}$, equiped with the norm $||.||_{k,p} = \Big(\sum\limits_{|\gamma|\leq k} ||D^\gamma.||_p^p\Big)^{1/p}$, is a Banach space. Here, and throughout this paper, the notation $||.||_p$ stands for the Lebesgue $\mathrm{L}^p(\mathbb{R}^n)$-norm and the differential operator D^γ stands for $D_x^\gamma = \frac{\partial^{\gamma_1}}{\partial x_1^{\gamma_1}}\frac{\partial^{\gamma_2}}{\partial x_2^{\gamma_2}} \cdots \frac{\partial^{\gamma_n}}{\partial x_n^{\gamma_n}}$ where $x = (x_1, x_2, \cdots, x_n) \in \mathbb{R}^n$. We studied, in [3], the equations $u_t + (L^*AL)u = a.\nabla^\theta(\psi(u))$, where $L = (-1)^m \sum\limits_{|\alpha|=|\beta|=m} a_{\alpha\beta}D^{\alpha+\beta}$ with nonnegative constant coefficients and A is a bounded nonnegative function independent of time. In particular, we derived the regularity in Sobolev spaces and the large time behavior for solutions. In [4], the authors considered the thin-film equations

$$\begin{cases} u_t + (-\Delta)^m u = \nabla . f(\nabla u) \\ u(0) = \varphi(x)\,,\ \left(\frac{\partial^i u}{\partial \nu^i}\right)_{\partial\Omega} = 0, \end{cases}$$

where $i \leq m-1$, $m \in \mathbb{N}$ $(m \geq 2)$ and Ω is a bounded domain. The main result they proved concerns the existence and the uniqueness of the solution to this problem. More precisely, they showed, under some conditions on the domain Ω and the function φ, the existence of a unique mild solution in $\mathcal{C}([0,\infty); L^p(\Omega))$ when the L^p-norm of φ is small enough.

In this work, we show that there is a constant $C > 0$ such that the solution u of (0.1) verifies, for all $t > 0$ and all $\gamma \in \mathbb{N}^n$ such that $0 \leq |\gamma| \leq m$, the estimates

$$\|D^\gamma u(t)\|_q \leq C\, t^{-(n(\frac{1}{p}-\frac{1}{q})+|\gamma|)/4m},$$

and the following large time behavior resulting from the nonlinear vector field $\mathcal{J}$,

$$\lim_{t\to\infty}\left(t^\alpha \left\| D^\gamma v(t) + \left(\iint_{\mathbb{R}^n\times[0,\infty]} J(\nabla^\mu u(z,s))dzds\right).\nabla^\theta(D^\gamma e^{-tL})\right\|_q\right) = 0$$

for all $\gamma \in \mathbb{N}^n$ such that $|\gamma| + \theta \leq 2m-1$ and, where $v(t) = u(t) - e^{-tL}u_0$ and $\alpha = \frac{n(1/p-1/q)+|\gamma|+\theta}{4m}$.

The solutions, considered for our problem, are mild L^p-solutions. More precisely, for $T \in (0,+\infty]$ and $u_0 \in L^p(\mathbb{R}^n)$, with $1 \leq p \leq \infty$, a solution u to the equation (0.1) is a function in $L^1_{loc}\big([0,T)\,,\, L^p(\mathbb{R}^n)\big)$ verifying the integral equation

$$u(t) = e^{-tL_0}u_0 + \int_0^t e^{-(t-s)L_0}\Big(\nabla^\theta.\mathcal{J}\big(\nabla^\mu u(s)\big)\Big)ds\,, \tag{1.1}$$

for all $t \in (0,T)$. A classical argument (see for example [2]), using an integration with respect to x and the Fourier transform, enables to show that the unique solution to (0.1) satisfies, for all $t \geq 0$, the conservation of mass $\int_{\mathbb{R}^n} u(x,t)dx = \int_{\mathbb{R}^n} u_0(x)dx$ and the L^τ-contraction property: for all $\tau \in [1,\infty]$,

$$||u(t)||_\tau \leq ||u_0||_\tau\,. \tag{1.2}$$

We recall, for all the sequel, that the heat kernel $\mathcal{K}_t : (x,t) \longmapsto \mathcal{K}(x,t)$ of the semigroup $e^{-t\mathcal{A}}$, associated with the operator $\mathcal{A} = \Delta^m a \Delta^m$, satisfies the following upper Gaussian bounds (see [5]): there exists a constant $C > 0$ such that for all $t > 0$,

$$||D^\gamma \mathcal{K}_t||_\tau \leq C\, t^{-(n(1-1/\tau)+|\gamma|)/4m} \tag{1.3}$$

for all $\tau \in [1,\infty]$ and for all multi-index $\gamma = (\gamma_1, \gamma_2, \cdots, \gamma_n) \in \mathbb{N}^n$ such that $|\gamma| = \gamma_1 + \gamma_2 + \cdots + \gamma_n \leq 2m - 1$.

2. Higher Order Gradient Estimates of Solutions

Before stating the main result of this section, let us mention that the classical Banach fixed point argument allows to establish that there exists a unique solution $u \in \mathcal{C}\big([0,\infty); L^1(\mathbb{R}^n)\big)$ to (0.1) such that $u \in \mathcal{C}\big((0,\infty); W^{4m,p}(\mathbb{R}^n)\big) \cap \mathcal{C}^1\big((0,\infty); L^p(\mathbb{R}^n)\big)$ for all $p \in (1,\infty)$. This solution satisfies the conservation of mass and the contraction property (1.2) (see for example [2], [6]).

The following result concerns the a priori L^q-estimates, for some $q > 0$, of the solution to the initial value problem (0.1). To prove these estimates, we need some technical tools. Among them the Gagliardo-Nirenberg inequality [7], which we recall for the reader's comfort: for the derivatives $D^j u$ such that $0 \leq j < m$, there exists a constant $C > 0$ (depending only on n, m, j, β, δ and τ) such that

$$||D^j u||_\alpha \leq C\, ||D^m u||_\delta^\tau ||u||_\beta^{1-\tau}, \tag{2.1}$$

where $1 \leq \beta, \delta \leq \infty$, $\tau \in [j/m, 1]$ and $1/\alpha = j/n + \tau(1/\delta - m/n) + (1-\tau)/\beta$.

Another tool which will be useful for our purpose is the following non classical version of Gronwall's lemma.

Lemma 1. *Let $0 \leq \beta < \alpha < 1$, $c > 0$, and f be a positive continuous function on $[0,T]$ such that for all $t \in]0,T]$,*

$$f(t) \leq c\Big(t^{-\alpha} + \int_0^t (t-s)^{-\beta} f(s)ds\Big). \tag{2.2}$$

Then, there exists a constant $C > 0$ such that for all $t \in]0,T]$,

$$f(t) \leq C\, t^{-\alpha}.$$

Proof. The proof of this inequality is an adaptation of the classical arguments used, for example, in [8], [9], [10].
According to the integral inequation (2.2), we have for all $t \in (0, T]$,

$$\begin{aligned} t^{\alpha} f(t) &\le c\Big(1 + t^{\alpha} \int_0^t (t-s)^{-\beta} s^{-\alpha} \left(s^{\alpha} f(s)\right) ds\Big) \\ &\le c\Big(1 + t^{\alpha}\Big(\int_0^{\nu t} \cdots + \int_{\nu t}^{t} \cdots\Big)\Big) \qquad \text{(where } 0 < \nu < 1) \\ &\le c\Big(1 + \epsilon\,\xi(t) + (1-\nu)^{-\beta} t^{\alpha-\beta} \int_0^{\nu t} s^{-\alpha} \left(s^{\alpha} f(s)\right) ds\Big), \end{aligned}$$

where $\xi(t) = \sup_{s\in[0,t]} \left(s^{\alpha} f(s)\right)$ and $\epsilon = (1-\nu)^{1-\beta} T^{1-\beta}/(1-\beta)\nu^{\alpha}$.
Now, by choosing ν such that the term $c\epsilon$ is small enough and by taking the supremum on $(0, \tau]$, we obtain for all $\tau \in [0, T)$, the classical integral inequation

$$\xi(\tau) \le k_1 + k_2 \int_0^{\tau} s^{-\alpha}\, \xi(s) ds, \tag{2.3}$$

where $k_1 = c/(1 - c\epsilon)$ and $k_2 = cT^{\alpha-\beta}/(1 - c\epsilon)(1-\nu)^{\beta}$. This implies, by the classical Gronwall's lemma, that

$$\xi(t) \le k_1 e^{k_2 T^{1-\alpha}/(1-\alpha)}$$

and ends the proof of Lemma 1. $\square$

Proposition 2. *Let $1 \le p \le q \le \infty$ and $u_0 \in L^p(\mathbb{R}^n)$. Let μ and θ be positive integers such that $\max(\mu, \theta) \le m - 1$ and $|\mathcal{J}(X)| \le C|X|^{\frac{mp+n}{\mu p+n}}$. Suppose that*

$$\begin{cases} (m+\theta)p < n < 3mp \text{ and } q > \frac{np}{n-(m+\theta)p} \\ \text{or} \\ n > 3mp \text{ and } \frac{np}{n-(m+\theta)p} < q < \frac{np}{n-3mp}. \end{cases}$$

Then there exists a constant $C > 0$ such that the solution u of (0.1) satisfies

$$\|D^{\gamma} u(t)\|_q \le C\, t^{-(n(\frac{1}{p}-\frac{1}{q})+|\gamma|)/4m},$$

for all $t > 0$ and all $\gamma \in \mathbb{N}^n$ such that $0 \le |\gamma| \le m$.

Proof. Let q be as above. According to the Duhamel integral equation (1.1), we have

$$\|\nabla^m u(t)\|_q \leq \|\nabla^m \mathcal{K}_t * u_0\|_q + \int_0^t \left\|\nabla^m \left(\nabla^\theta . (\mathcal{K}(t-s) * \mathcal{J}(\nabla^\mu u(s)))\right)\right\|_q ds.$$

It then follows by (1.3) and Young's inequality the following estimates

$$\begin{aligned}\|\nabla^m u(t)\|_q &\leq C t^{-\frac{n}{4m}(1/p-1/q)-1/4}\|u_0\|_p + \int_0^t (t-s)^{-(m+\theta)/4m}||\mathcal{J}(\nabla^\mu u(s))||_q ds \\ &\leq C t^{-\frac{n}{4m}(1/p-1/q)-1/4}\|u_0\|_p + \int_0^t (t-s)^{-(m+\theta)/4m}||\nabla^\mu u(s)||_{q\omega}^\omega \, ds\,,\end{aligned}$$

where $\omega = \frac{mp+n}{\mu p+n}$. Here, and throughout this paper, generic constants are denoted by C.

Since $\mu < m$, then by using the Gagliardo-Nirenberg inequality (2.1) we obtain

$$||\nabla^\mu u||_{q\omega} \leq C\,||\nabla^m u||_q^{1/\omega}||u||_p^{(\omega-1)/\omega},$$

which implies the following integral inequation

$$\begin{aligned}\|\nabla^m u(t)\|_q &\leq C t^{-\frac{n}{4m}(1/p-1/q)-1/4}\|u_0\|_p + \int_0^t (t-s)^{-(m+\theta)/4m}||\nabla^m u(s)||_q||u(s)||_p^{\omega-1}\, ds \\ &\leq C t^{-\frac{n}{4m}(1/p-1/q)-1/4}\|u_0\|_p + ||u_0||_p^{\omega-1}\int_0^t (t-s)^{-(m+\theta)/4m}||\nabla^m u(s)||_q\, ds\end{aligned}$$

thanks to the contraction property (1.2).

Now, taking into account the conditions on the parameters, Lemma 1 implies that for all $t > 0$,

$$\|\nabla^m u(t)\|_q \leq C t^{-\frac{1}{4m}\left(n(\frac{1}{p}-\frac{1}{q})+m\right)}. \tag{2.4}$$

On the other hand, using (2.4) and the Sobolev embedding theorem, with $\delta = \frac{nq}{n+mq} > 1$, we obtain that

$$\begin{aligned}||u(t)||_q &\leq C\,||\nabla^m u(t)||_\delta \\ &\leq C t^{-\frac{n}{4m}(\frac{1}{p}-\frac{1}{q})}.\end{aligned}$$

Finally, this estimate and (2.4) imply, for all $\gamma \in \mathbb{N}^n$ such that $0 < |\gamma| < m$,

$$\begin{aligned}||D^\gamma u(t)||_q &\leq C\,||\nabla^m u(t)||_q^{|\gamma|/m}\,||u(t)||_q^{1-|\gamma|/m} \\ &\leq C t^{-\frac{1}{4m}\left(n(\frac{1}{p}-\frac{1}{q})+|\gamma|\right)}.\end{aligned}$$

3. Large Time Behavior Related to Nonlinear Effects

In this section, we consider sharp L^q-estimates for derivatives of solutions to (0.1). These estimates are considered more precise than those given in Proposition 2. We suppose that there exists a constant $C > 0$ such that the solution u of (0.1) satisfies

$$\|D^\gamma u(t)\|_q \leq C\,(1+t)^{-(n(\frac{1}{p}-\frac{1}{q})+|\gamma|)/4m}, \tag{3.1}$$

for all $t > 0$ and all $\gamma \in \mathbb{N}^n$ such that $0 \leq |\gamma| \leq m$. These sharp estimates are valid when, for example, $u_0 \in L^1(\mathbb{R}^n) \cap L^\infty(\mathbb{R}^n)$ (see [2]-Corollary 3.2, [6]).

3.1. Statement of the Main Result

Our objective is to investigate the large time asymptotics for solutions to the Cauchy problem for higher order nonlinear dissipative equations (0.1). More precisely, we show the following behavior of solutions.

Theorem 3. *Suppose that* $|J(X)| \leq C|X|^\omega$ *with* $\omega > \frac{(n+4m)p}{n+p\mu}$. *Then the solution of (0.1) verifies*

$$t^{(n(1/p-1/q)+|\gamma|+\theta)/4m}\times$$

$$\left\| D^\gamma(u(t)-e^{-tL}u_0) + \Big(\iint_{\mathbb{R}^n\times[0,\infty]} J(\nabla^\mu u(z,s))dzds\Big).\nabla^\theta(D^\gamma e^{-tL})\right\|_q \xrightarrow[t\to\infty]{} 0 \tag{3.2}$$

for all $\gamma \in \mathbb{N}^n$ *such that* $|\gamma| + \theta \leq 2m-1$.

Before proving this result, let us mention that we have the following asymptotic behavior of solutions to (0.1) as for the heat equation. Under the hypotheses of Theorem 3, there exists a constant $C > 0$ such that

$$\left\|D^\gamma u(t) - \Big(\int_{\mathbb{R}^n} u_0(x)dx\Big)\, D^\gamma K(t)\right\|_q \leq C\,t^{-(n(1/p-1/q)+|\gamma|+1)/4m}.$$

For the reader's convenience, we sketch out the computations for u_0 in the weighted space $L^1(\mathbb{R}^n, 1+|x|) = \{f \in L^1(\mathbb{R}^n)/ \int_{\mathbb{R}^n}(1+|x|)|f(x)|dx < \infty\}$

which is dense into $L^1(\mathbb{R}^n)$. Put $m_0 = \displaystyle\int_{\mathbb{R}^n} u_0(x)dx$ and then

$$\|D^\gamma u(t) - m_0\, D^\gamma K(t)\|_q \leq$$

$$\underbrace{\|D^\gamma K(t) * u_0 - m_0\, D^\gamma K(t)\|_q}_{I_1(t)} + \underbrace{\int_0^t \left\|\nabla^\theta.\big(D^\gamma \mathcal{K}(t-s) * J\big(\nabla^\mu u(s)\big)\big)\right\|_q ds}_{I_2(t)}$$

The well known estimates on the heat kernel (see [13]) yield

$$I_1(t) \leq Ct^{-(n(1/p-1/q)+|\gamma|)/4m-1/4m}$$

and by applying Young's inequality and (3.1), we obtain

$$\begin{aligned} I_2(t) \leq C\Big(&\int_0^{t/2} (t-s)^{-n(1-1/q)/4m-(|\gamma|+\theta)/4m} ||\nabla^\mu u(s)||_\omega^\omega ds \\ &+ \int_{t/2}^t (t-s)^{-(|\gamma|+\theta)/4m} ||\nabla^\mu u(s)||_{q\omega}^\omega\, ds\Big) \\ \leq C\Big(&\int_0^{t/2} (t-s)^{-n(1-1/q)/4m-(|\gamma|+\theta)/4m} (1+s)^{-\omega(n(1/p-1/\omega)+\mu)/4m} ds \\ &+ \int_{t/2}^t (t-s)^{-(|\gamma|+\theta)/4m} (1+s)^{-\omega(n(1/p-1/q\omega)+\mu)/4m} ds\Big). \end{aligned}$$

Then, taking into account the fact that $\omega > \frac{(n+4m)p}{n+p\mu}$, a simple computation yields

$$I_2(t) \leq C\,(t/2)^{-(n(1/p-1/q)+|\gamma|)/4m} t^{-1/4m}\,.$$

3.2. Proof of Theorem 3

According to the integral equation (1.1), we will estimate the L^q-norm of the term

$$\xi(t) = \underbrace{\Big(\iint_{\mathbb{R}^n\times[0,\infty]} J(\nabla^\mu u(z,s))dzds\Big).\nabla^\theta(D^\gamma e^{-tL})}_{\xi_1(0,\infty)} - \underbrace{\int_0^t \nabla^\theta.[(D^\gamma e^{-(t-s)L})J(\nabla^\mu u(s))]\, ds}_{\xi_2(0,t)}.$$

We then use the following decomposition

$$||\xi(t)||_q \leq ||\xi_1(t/2,\infty)||_q + ||\xi_2(t/2,t)||_q + ||\xi_1(0,t/2) - \xi_2(0,t/2)||_q.$$

First of all, we have

$$||\xi_1(t/2,\infty)||_q \;\leq\; Ct^{-n(1-1/q)/4m-(|\gamma|+\theta)/4m} \underbrace{\int_{t/2}^\infty \int_{\mathbb{R}^n} |J(\nabla^\mu u(z,s))| dzds}_{\mathcal{R}(t)}$$

and due to (3.1), we have $\mathcal{R}(t) \leq \int_{t/2}^{\infty} ||\nabla^{\mu} u(s)||_{\omega}^{\omega}\, ds \leq \int_{t/2}^{\infty} (1+s)^{-\omega(n(1/p-1/\omega)+\mu)/4m} ds$ which converges under the condition $\omega > \frac{(n+4m)p}{n+p\mu}$. Therefore, since the term $\left(t^{-n(1-1/p)/4m}\mathcal{R}(t)\right)$ goes to 0 when $t \to \infty$, we obtain

$$\lim_{t\to\infty}\left(t^{(n(1/p-1/q)+|\gamma|+\theta)/4m}||\xi_1(t/2,\infty)||_q\right) = 0.$$

To estimate the second term, we use Young's inequality, the estimates on the derivatives of the heat kernel and (3.1) to obtain

$$\begin{aligned}
||\xi_2(t/2,t)||_q &\leq C\int_{t/2}^{t}(t-s)^{-(|\gamma|+\theta)/4m}||J(\nabla^{\mu}u(s))||_q\, ds\\
&\leq C\int_{t/2}^{t}(t-s)^{-(|\gamma|+\theta)/4m}||\nabla^{\mu}u(s)||_{q\omega}^{\omega}\, ds\\
&\leq C\int_{t/2}^{t}(t-s)^{-(|\gamma|+\theta)/4m}(1+s)^{-\omega(n(1/p-1/q\omega)+\mu)/4m}\, ds\\
&\leq Ct^{-(n(1/p-1/q)+|\gamma|+\theta)/4m}t^{-\tau},
\end{aligned}$$

where $\tau = ((n+p\mu)\omega - 4mp - n)/4mp > 0$ since $\omega > \frac{(n+4m)p}{n+p\mu}$. It then follows that

$$\lim_{t\to\infty}\left(t^{n(1/p-1/q)/4m+(|\gamma|+\theta)/4m}||\xi_2(t/2,t)||_q\right) = 0.$$

To estimate the third term $||\xi_1(0,t/2)-\xi_2(0,t/2)||_q$, we write

$$\xi_2(0,t/2)-\xi_1(0,t/2) = \iint_{D_t}\nabla^{\theta}\left(D^{\gamma}K(.-z,t-s)-D^{\gamma}K(.,t)\right).J(\nabla^{\mu}u(z,s))dzds,$$

where $D_t = \mathbb{R}^n\times[0,t/2]$, and we set up the estimations for $q=1$ and $q\to\infty$ the case $1<q<\infty$ then follows by interpolation.

For the case $q=1$, we have

$$||\xi_2(0,t/2)-\xi_1(0,t/2)||_1 \leq \underbrace{||\nabla^{\theta}(D^{\gamma}K(t))||_1}_{\leq ct^{-(|\gamma|+\theta)/4m}}\underbrace{\iint_{\mathbb{R}^n\times[0,\infty]}|J(\nabla^{\mu}u(z,s))|dzds}_{<\infty\ (\text{since } \omega>\frac{(n+4m)p}{n+p\mu})} +$$

$$+\int_0^{t/2}\underbrace{||\nabla^\theta(D^\gamma K(t-s))||_1}_{\le c(t-s)^{-(|\gamma|+\theta)/4m}}\underbrace{||J(\nabla^\mu u(s))||_1}_{\le c(1+s)^{-\frac{\omega}{4m}(n(\frac{1}{p}-\frac{1}{\omega})+\mu)}}ds$$

$$\le C(t/2)^{-n(1/p-1)/4m-(|\gamma|+\theta)/4m}\rho(t),$$

where the function $t\longmapsto\rho(t)$ is such that

$$\rho(t)=t^{-n(1-1/p)/4m}\Big(1+\int_0^{t/2}(1+s)^{-\omega(n(1/p-1/\omega)+\mu)/4m}ds\Big)\begin{cases}\xrightarrow[t\to\infty]{} 0 \text{ if } p>1\\ \xrightarrow[t\to\infty]{} c \text{ if } p=1\end{cases}$$

and the constant c is nonnegative.

For the case when $q\to\infty$, write

$$||\xi_2(0,t/2)-\xi_1(0,t/2)||_\infty\le\iint_D||\cdots||_\infty\,dzds+\iint_{D'}||\cdots||_\infty\,dzds,$$

where $D=\{|z|\le(\frac{t}{r})^{1/4m}\}\times[0,(\frac{t^{4m}}{r})^{1/4m}]$, for a fixed r large enough, and D' is the supplementary subset of D into $D_t=\mathbb{R}^n\times[0,t/2]$.
As done before, the estimates on the derivatives of the heat kernel imply

$$\iint_{D'}||\cdots||_\infty\,dzds\le$$

$$\le\iint_{D'}\Big(\underbrace{||\nabla^\theta(D^\gamma K(.-z,t-s))||_\infty}_{\le c(t-s)^{-(n+|\gamma|+\theta)/4m}}+\underbrace{||\nabla^\theta(D^\gamma K(.,t))||_\infty}_{\le ct^{-(n+|\gamma|+\theta)/4m}}\Big)|J(\nabla^\mu u(z,s))|dzds$$

$$\le C(t/2)^{-(n+(|\gamma|+\theta)p)/4mp}\rho_1(t),$$

where $\rho_1(t)=t^{-n(1-1/p)/4m}\int_{t/r^{1/4m}}^{t/2}(1+s)^{-\omega(n(1/p-1/\omega)+\mu)/4m}ds\xrightarrow[t\to\infty]{}0$ since $\omega(n(1/p-1/\omega)+\mu)/4m>1$ and then

$$\lim_{t\to\infty}\Big(t^{-(n+(|\gamma|+\theta)p)/4mp}\iint_{D'}||\cdots||_\infty\,dzds\Big)=0.\qquad(3.3)$$

For the term $\iint_D ||\cdots||_\infty dzds$, we have

$$\begin{aligned}
&\iint_D ||\cdots||_\infty dzds \leq \\
&\leq \sup_{(z,s)\in D} \Big\| \nabla^\theta \Big(D^\gamma \big(K(.-z,t-s) - K(.,t) \big) \Big) \Big\|_\infty \underbrace{\iint_D |J(\nabla^\mu u(z,s))| dzds}_{\rho_2(t)} \\
&\leq \Big(\sup_{0<s\leq t/r^{1/4m}} \Big\| \nabla^\theta \Big(D^\gamma \big(K(.,t-s) - K(.,t) \big) \Big) \Big\|_\infty + \\
&\qquad \sup_{|z|\leq (t/r)^{1/4m}} \Big\| \nabla^\theta \Big(D^\gamma \big(K(.-z,t) - K(.,t) \big) \Big) \Big\|_\infty \Big) \rho_2(t) \\
&:= \big(\delta_1(t) + \delta_2(t)\big)\rho_2(t).
\end{aligned}$$

Now, by using the inverse Fourier transform, we obtain

$$\delta_1(t) \leq C \sup_{0<s\leq t/r^{1/4m}} \Big\| \int_{\mathbb{R}^n} (i\nu)^{|\gamma|+\theta} \big(e^{-a(t-s)|\nu|^{4m}} - e^{-at|\nu|^{4m}} \big) e^{ix\nu} d\nu \Big\|_\infty,$$

where $\widehat{e^{-tL}f}(\nu) = H(\nu)\hat{f}(\nu)$ with $a|\nu|^{4m} \leq H(\nu) \leq b|\nu|^{4m}$ for two nonnegative constants a and b. Taking into account the fact that $||\hat{f}||_\infty \leq ||f||_1$ and using the mean value theorem yield

$$\begin{aligned}
\delta_1(t) &\leq C \sup_{0<s\leq t/r^{1/4m}} \int_{\mathbb{R}^n} |\nu|^{|\gamma|+\theta} \big| e^{-a(t-s)|\nu|^{4m}} - e^{-at|\nu|^{4m}} \big| d\nu \\
&\leq C \sup_{0<s\leq t/r^{1/4m}} \int_{\mathbb{R}^n} |\nu|^{|\gamma|+\theta} |\nu|^{4m} (as) e^{-a(t-s)|\nu|^{4m}} d\nu \\
&\leq C(t/r^{1/4m}) \int_{\mathbb{R}^n} |\nu|^{4m+|\gamma|+\theta} e^{-a(r^{1/4m}-1)t|\nu|^{4m}/r^{1/4m}} d\nu \\
&\leq \Big(\frac{C}{r^{1/4m}} \Big) t^{1-(n+4m+|\gamma|+\theta)/4m} \int_{\mathbb{R}^n} |\nu|^{4m+|\gamma|+\theta} e^{-a(r^{1/4m}-1)|\nu|^{4m}/r^{1/4m}} d\nu,
\end{aligned}$$

which involves

$$t^{(n+(|\gamma|+\theta)p)/4mp} \delta_1(t) \leq C \Big(\frac{t^{-n(1-1/p)/4m}}{r^{1/4m}} \Big). \tag{3.4}$$

As for $t \longmapsto \delta_1(t)$ and by using the "time-shifting" property of the Fourier transform, we obtain

$$\begin{aligned}
\delta_2(t) &\leq C \sup_{|z|\leq (t/r)^{1/4m}} \int_{\mathbb{R}^n} |\nu|^{|\gamma|+\theta} \big| e^{-iz\nu} - 1 \big| e^{-at|\nu|^{4m}} \, d\nu \\
&\leq C \sup_{|z|\leq (t/r)^{1/4m}} \int_{\mathbb{R}^n} |\nu|^{|\gamma|+\theta+1} |z| e^{-at|\nu|^{4m}} \, d\nu \\
&\leq \Big(\frac{C}{r^{1/4m}} \Big) t^{1/4m-(n+|\gamma|+\theta+1)/4m} \int_{\mathbb{R}^n} |\nu|^{|\gamma|+\theta+1} e^{-a|\nu|^{4m}} \, d\nu,
\end{aligned}$$

which means, as for $\delta_1(t)$, that

$$t^{(n+(|\gamma|+\theta)p)/4mp} \delta_2(t) \leq C \Big(\frac{t^{-n(1-1/p)/4m}}{r^{1/4m}} \Big). \tag{3.5}$$

Finally, by combining (3.4), (3.5) and by taking r large enough, we get

$$\lim_{t\to\infty} \Big(t^{(n+(|\gamma|+\theta)p)/4mp} \iint_D || \cdots ||_\infty \, dz ds \Big) = 0$$

and then, in view of (3.3), we obtain

$$\Big(t^{(n+(|\gamma|+\theta)p)/4mp} ||\xi_2(0,t/2) - \xi_1(0,t/2)||_\infty \Big) \xrightarrow[t\to\infty]{} 0\,,$$

thanks to the inequality

$$||\xi_2(0,t/2) - \xi_1(0,t/2)||_\infty \leq C t^{-(n+(|\gamma|+\theta)p)/4mp} \Big(\rho_1(t) + t^{-\frac{n}{4m}(1-\frac{1}{p})} \rho_2(t)/r^{1/4m} \Big).$$

The estimates on $||\xi_2(0,t/2) - \xi_1(0,t/2)||_1$ obtained above and the last result allow to conclude the proof of theorem 3 by interpolation.

REFERENCES

[1] Carlen, E.A. and Loss, M. (1995). Optimal smoothing and decay estimates for viscously damped conservation laws, with applications to the 2-D Navier-Stokes equation, *Duke Math. J.* 81 (1): 135-157.

[2] Biler, P., Karch, G. and Woyczyński, W.A. (2001). Asymptotics for conservation laws involving Lévy diffusion generators, *Studia Mathematica* 148: 171-192.

[3] Qafsaoui, M. (2006). Large time behavior for nonlinear higher order convection-diffusion equations, *Ann. Inst. H. Poincaré Anal. Non Linéaire*, 23: 911-927.

[4] Sandjo, A.N. and Soh, C.W. (2014). Space-time estimates of mild solutions of a class of higher-order semilinear parabolic equations in L^p, *Nonautonomous Dynamical Systems*, 1: 83-92.

[5] Auscher, P. and Qafsaoui, M. (2000). Equivalence between regularity theorems and heat kernel estimates for higher order elliptic operators and systems under divergence form, *J. Funt. Anal.* 177: 310-364.

[6] Escobedo, M. and Zuazua, E. (1991). Large time behavior for convection-diffusion equations in $\mathbb{R}^n$, *J. Funt. Anal.* 100 (1): 119-161.

[7] Nirenberg, L. (1959). On elliptic partial differential equations, *Annali della Scuola Norm. Sup. Pisa*, 13: 115-162.

[8] Bidégaray, B. and Moisan, L. (2000). Petits problèmes de mathématiques appliquées et de modélisation [Some applied mathematics an modelisation problems], *Scopos*, vol. 9, Springer.

[9] Cazenave, T. and Haraux, A. (1990). *Introduction aux problèmes d'évolution semi-linéaires [Introduction to semilinear evolution problems]*, Ellipses.

[10] Henry, D. (1981). Geometric theory of semilinear parabolic equations, *Lecture Note in Math*. Springer, 840.

[11] Bae, H.-O. and Jin, B.J. (2006). Asymptotic behavior for the Navier-Stokes equations in 2D exterior domains, *J. Funt. Anal.* 240 (2): 508-529.

[12] Friedman, A. (1983). *Partial differential equations*, R.E. Krieger Publishing Company.

[13] Kirane, M. and Qafsaoui, M. (2002). On the asymptotic behavior for convection-diffusion equations associated to higher order elliptic operators under divergence form, *Rev. Mat. Complut.* 15 (2): 585-598.

[14] Qin, Y. (2017). Analytic inequalities and their applications in PDEs, *Advances in Partial Differential Equations*, Birkhäuser.

[15] Shlesinger, M.F., Zaslavsky, G.M. and Frisch, U. (2013). Lévy flights and related topics in physics, *Lecture Notes in Physics*, Springer, 450.

[16] Zuazua, E. (1993). Weakly nonlinear large time behavior in scalar convection-diffusion equations, *Differential Integral Equations*, 6: 1481-1491.

In: Evolutionary Processes and Applications
Editor: G. M. N'Guérékata
ISBN: 978-1-53616-350-6

Chapter 6

DISCRIMINATING SENTINEL WITH GIVEN SENSITIVITY FOR THE HEAT EQUATION WITH FOURIER BOUNDARY CONDITION

Sadou Tao [*] ***and Elisée Gouba*** [†]
Université Joseph Ki-Zerbo, Département de mathématiques,
Ouagadougou, Burkina Faso

Abstract

In this article, we study the null controlability problem with two constraints on a pair of controls. We apply these results to a discriminating sentinel with given sensitivity to detect some parameters in a pollution problem governed by a semi-linear parabolic equation with Fourier boundary condition.

Keywords: systems governed by PDEs, nonlinear PDEs of parabolic type, null controllability, Carleman inequalities, discriminating sentinel with given sensitivity

1. INTRODUCTION

For $N \in \mathbb{N}^*$, let Ω be an open bounded subset of $\mathbf{R}^N$ of boundary Γ of class $\mathcal{C}^2$. Let also ω be an non-empty open bounded subset of Ω. For a time $T > 0$,

[*] Corresponding Author's Email: sadoutao@yahoo.fr.
[†] Corresponding Author's Email: elgouba@gmail.com.

we set $Q = \Omega \times (0,T)$, $\Sigma = \Gamma \times (0,T)$ and $U = \omega \times (0,T)$. We consider the following parabolic equation:

$$\begin{cases} -\dfrac{\partial q}{\partial t} - \Delta q + a_0 q &= h + v\chi_\omega + w\chi_\omega \quad \text{in} \quad Q \\ \dfrac{\partial q}{\partial \nu} + bq &= 0 \quad \text{on} \quad \Sigma \\ q(T) &= 0 \quad \text{in} \quad \Omega \end{cases} \tag{1}$$

where $a_0 \in L^\infty(Q)$, $h \in L^2(Q)$ and $b \in L^\infty(\Sigma)$ with $b \geq 0$. The controls v and w belong to $L^2(U)$, χ_ω denotes the characteristic function of ω. It is well know that the problem (1) admits a unique solution q in the Hilbert space

$$L^2(0,T,H^1(\Omega)) \cap C\left((0,T), L^2(\Omega)\right)$$

and $\dfrac{\partial q}{\partial t} \in L^2(0,T,H^{-1}(\Omega))$ (we refer to [6])

Let

$$Y_\lambda \text{ and } \mathcal{M} \text{ be two } \mathit{real} \text{ closed vector subspaces of } L^2(U) \tag{2}$$

$Y_\lambda^\perp$ and $\mathcal{M}^\perp$ respectively the orthogonal subspaces of Y_λ and $\mathcal{M}$ in $L^2(U)$.

We use the following notation

$$q = q(x,\ t;\ (v,w)).$$

This mean that the solution q of (1) depend for two controls v and w.

The problem is the following:

Let a function h be in $L^2(Q)$, $a_0 \in L^\infty(Q)$ find two controls $\widehat{v}$ and $\widehat{w}$ in $L^2(U)$ with

$$\widehat{v} \in Y_\lambda^\perp \text{ and } \widehat{w} \in \mathcal{M}^\perp \tag{3}$$

such that if $q = q(x,t;(\widehat{v},\widehat{w}))$ is the unique solution of

$$\begin{cases} -\dfrac{\partial q}{\partial t} - \Delta q + a_0 q &= h + \widehat{v}\chi_\omega + \widehat{w}\chi_\omega \quad \text{in} \quad Q \\ \dfrac{\partial q}{\partial \nu} + bq &= 0 \quad \text{on} \quad \Sigma \\ q(T) &= 0 \quad \text{in} \quad \Omega \end{cases} \tag{4}$$

then

$$q(.,\ 0;\ (\widehat{v},\widehat{w})) = 0 \text{ in } \Omega \tag{5}$$

and the couple $(\widehat{v}, \widehat{w})$ is the minimal norm in $L^2(U)$. i.e.,

$$||(\widehat{v}, \widehat{w})|| = \min_{(v,w)\in\mathcal{E}} ||(v, w)|| \tag{6}$$

where

$$||(v, w)|| = (||v||^2_{L^2(U)} + ||w||^2_{L^2(U)})^{\frac{1}{2}}$$

and

$$\mathcal{E} = \left\{ \begin{array}{c} (v, w) \in Y_\lambda^\perp \times \mathcal{M}^\perp, \text{ such that the pair } ((v, w),\ q(x, t, (v, w))) \\ \text{satisfies (3) and (5)} \end{array} \right\}.$$

The problem (3)-(5) is a null-controllability problem with constraint on the control. If $Y_\lambda^\perp = \{0\}$ and $M^\perp = \{0\}$, this problem becomes a null-controllability problem without constraints on the control. There are a large literature on this problem. In [13] G. Lebeau and L. Robbiano solved this problem when $a_0 \equiv 0$ and $b \equiv 0$. E. Fernandez-Cara, M. González-Burgos, S. Guerrero and J.P. Puel in [8] used the Carleman estimate for the weak solution of heat equation with non homogeneous Neumann boundary conditions to prove the null-controllability of the linear heat equation. In the nonlinear case A. Fursikov and O. Yu. Imanuvilov in [2], [21] showed using a Carleman's estimate that, when the control acts on the boundary, null controllability holds for bounded continuous and sufficiently small initial data. One can also see [23, 1, 9, 10, 12], [5], [6], [7] and the references in them. If $Y_\lambda^\perp \neq \{0\}$ or $\mathcal{M}^\perp \neq \{0\}$ this problem becomes a null controllabilty problem with one constraint on the control, one can see [20, 14].

In this paper, we studied the case where $Y_\lambda^\perp \neq \{0\}$ and $\mathcal{M}^\perp \neq \{0\}$ showing that the null controllability problem (3)-(6) holds when $b \neq 0$.This required the establishment of a Carleman inequality with a Fourier boundary condition. Then we apply the result to discriminating sentinels with given sensitivity and with Fourier boundary condition

The main result of this paper is the follows:

We assume that

$$Y_\lambda^\perp \text{ and } \mathcal{M}^\perp \text{ are finite dimensional} \tag{7}$$

and

$$\left\{ \begin{array}{l} (\forall \rho \in Y_\lambda), (\dfrac{\partial \rho}{\partial t} - \Delta\rho + a_0 = 0 \text{ in } \omega \times (0, T) \Longrightarrow \rho = 0 \text{ in } \omega \times (0, T)) \\ (\forall \rho \in \mathcal{M}), (\dfrac{\partial \rho}{\partial t} - \Delta\rho + a_0 = 0 \text{ in } \omega \times (0, T) \Longrightarrow \rho = 0 \text{ in } \omega \times (0, T)) \end{array} \right. \tag{8}$$

Theorem 1. *We assume that (7), (8) are satisfied. There exist then the weight function* θ *(not vanish with* $\frac{1}{\theta}$ *bounded) such that for any function* $h \in L^2(Q)$ *with* $\theta h \in L^2(Q)$*, the null controlability problem (3)-(5) admits a solution* $(\widehat{v}_\theta, \widehat{w}_\theta)$*. Moreover, the pair* $(\widehat{v}_\theta, \widehat{w}_\theta)$ *is such that*

$$\begin{aligned} \widehat{v}_\theta &= -(\widehat{\rho}_\theta \chi_\omega - P_1 \widehat{\rho}_\theta) \text{ in } \omega \times (0,T) \\ \widehat{w}_\theta &= -(\widehat{\rho}_\theta \chi_\omega - P_2 \widehat{\rho}_\theta) \text{ in } \omega \times (0,T) \end{aligned} \tag{9}$$

where P_1 *and* P_2 *are respectively the orthogonal projection operator from* $L^2(\omega \times (0,T))$ *into* Y_λ *and into* $\mathcal{M}$*, and* $\widehat{\rho}_\theta$ *satisfies*

$$\left\{ \begin{aligned} \frac{\partial \widehat{\rho}_\theta}{\partial t} - \Delta \widehat{\rho}_\theta + a_0 \widehat{\rho}_\theta &= 0 \quad \text{in } Q \\ \frac{\partial \widehat{\rho}_\theta}{\partial \nu} + b \widehat{\rho}_\theta &= 0 \quad \text{on } \Sigma. \end{aligned} \right. \tag{10}$$

The rest of this paper is organized as follow. In the section 2 we prove the theorem 1 after established a adapted Carleman inequality to (4). In the section 3, we show by the penalization method the existence of solution of the controlability problem (3)-(5) of theorem 1. In the end, in the section 4 we give an application of these results at a sentinel, that will be define.

2. Null-Controlability with Two Constraints on the Control

2.1. Adapted Carleman Inequality

In order to state Carleman's inequality, we introduced now some objects and notations. We start by choosing the auxiliary function $\psi \in C^2(\overline{\Omega})$ which satisfies the following conditions:

$$\psi(x) > 0 \; \forall x \in \Omega, \; \psi(x) = 0 \; \forall x \in \Gamma, \; |\nabla \psi(x)| \neq 0 \; \forall x \in \overline{\Omega - \omega}.$$

The function ψ exists according to A. Fursikov and O. Imanuvilov in [2].

We define for any positive parameter value λ the following weight functions

$$\varphi(x,t) = \frac{e^{\lambda \psi(x)}}{t(T-t)}, \; \eta(x,t) = \frac{e^{2\lambda ||\psi||_\infty} - e^{\lambda \psi(x)}}{t(T-t)}.$$

we adopt the following notations

$$\begin{cases} L_0 & = & \dfrac{\partial}{\partial t} - \Delta \\ L & = & \dfrac{\partial}{\partial t} - \Delta + a_0 I \\ \mathcal{V} & = & \left\{ \rho \in C^\infty(\overline{Q}) / \dfrac{\partial \rho}{\partial \nu} + b\rho = 0 \text{ on } \Sigma \right\} \end{cases} \tag{11}$$

where $a_0 \in L^\infty(Q)$ and $b \in L^\infty(\Sigma), b \geq 0$ are defined in (1). The Carleman's inequality can be formulate by:

Proposition 2. *There exists three constants* $\lambda_1 = \lambda_1(\Omega, \omega) > 1$, $s_1 = s_1(\Omega, \omega, T) > 1$ *and* $C = C(\Omega, \omega, a_0)$ *such that for any* $\lambda \geq \lambda_1$, *any* $s \geq s_1$ *and any* $\rho \in \mathcal{V}$ *the following inequality holds*

$$\begin{aligned} & \int_0^T \int_\Omega e^{-2s\eta}((s\varphi)^{-1}(|\rho'|^2 + |\Delta\rho|^2) + s\lambda^2\varphi|\nabla\rho|^2 \\ & +s^3\lambda^4\varphi^3|\rho|^2)dxdt \\ \leq \; & C(\int_0^T \int_\Omega e^{-2s\eta}||L\rho||^2 dxdt + s^3\lambda^4 \int_0^T \int_\omega e^{-2s\eta}\varphi^3|\rho|^2 dxdt). \end{aligned} \tag{12}$$

There exists a weight function $\theta \in C^2(Q)$ such that $\dfrac{1}{\theta}$ bounded, and a constant $c = c(\Omega, a_0, b, T)$ such for any $\rho \in \mathcal{V}$, the following inequality holds

$$\int_0^T \int_\Omega \frac{1}{\theta^2} |\rho|^2 \, dxdt \leq C \left(\int_0^T \int_\Omega |L\rho|^2 \, dxdt + \int_0^T \int_\omega |\rho|^2 \, dxdt \right). \tag{13}$$

All these results are understood. We refer to E. Fernández-Cara and E. Zuazua in [11] and O. Nakoulima in [20].

In order to deal with the constraint (3) we have to derive a more precise observability inequality adapted to the spaces Y_λ and $\mathcal{M}$ defined in (2). The follow lemma is the key of our results.

Lemma 3. *Assume that (7) and (8) hold, then there exists a positive contant* C=$C(\Omega, \omega, b, a_0)$ *such that for any* $\rho \in \mathcal{V}$

$$\begin{aligned} & \int_0^T \int_\Omega \frac{1}{\theta^2} |\rho|^2 \, dxdt \\ \leq \; & C \left(\int_0^T \int_\Omega |L\rho|^2 \, dxdt + \int_0^T \int_\omega |\rho - P_1\rho|^2 \, dxdt + \int_0^T \int_\omega |\rho - P_2\rho|^2 \, dxdt \right) \end{aligned} \tag{14}$$

where P_1 and P_2 are respectivly the orthogonal projection from $L^2(\omega \times (0,T))$ into Y_λ and into $\mathcal{M}$.

2.2. Proof of Theorem1

The main tool is the observability inequality (14) adapted to the constraints.

Consider now the following symmetric bilinear form

$$a(\rho, \widehat{\rho}) = \int_Q L\rho L\widehat{\rho}dxdt + \int_{L^2(\omega\times(0,T))} (\rho - P_1\rho)(\widehat{\rho} - P_1\widehat{\rho})dxdt \quad (15)$$
$$+ \int_{L^2(\omega\times(0,T))} (\rho - P_2\rho)(\widehat{\rho} - P_2\widehat{\rho})dxdt.$$

Due to Lemma 3, this bilinear form is a scalar product on $\mathcal{V}$. Let V be the Hilbert space obtained upon taking the closure of $\mathcal{V}$ under the norm

$$\rho \longmapsto ||\rho||_V = \sqrt{a(\rho,\rho)}. \quad (16)$$

Remark 4. *Observe that the norm $\|.\|_V$ is related to the right-hand side of inequality (14). Similary, the left-hand side of (14) leads to the norm*

$$||\rho||_\theta = (\int_Q \frac{1}{\theta^2}|\rho|^2 dxdt)^{\frac{1}{2}}. \quad (17)$$

The completion of $\mathcal{V}$ is the weighted Hilbert space usually denoted by $L^2_{\frac{1}{\theta}}(Q)$.

The inequality (14) shows that

$$||\rho||_\theta \leq C||\rho||_V. \quad (18)$$

This inequality extends to $\rho \in V$. This shows that V is continuously imbedded in $L^2_{\frac{1}{\theta}}(Q)$. Let us now consider $h \in L^2(Q)$ such that $\theta h \in L^2(Q)$. Then due to (14) and the Cauchy-Schwartz inequality, we deduce that the linear form defined on V by:

$$\rho \longmapsto \int_Q h\rho dxdt$$

is continuous. By the Lax-Milgram theorem, for any $h \in L^2(Q)$ such that $\theta h \in L^2(Q)$ there exists one and only one solution $\rho_\theta \in V$ to the variational problem:

$$\forall \rho \in V \;\; a(\rho_\theta, \rho) = \int_Q h\rho dxdt \tag{19}$$

Proposition 5. *Assume that (7) and (8) hold. Let h be in $L^2(Q)$ such that $\theta h \in L^2(Q)$. Let ρ_θ be the unique solution of (19), $P_1\rho_\theta$ the projection from $\rho_\theta\chi_\omega$ into Y_λ and $P_2\rho_\theta$ the projection from $\rho_\theta\chi_\omega$ into $\mathcal{M}$. Set*

$$v_\theta = -(\rho_\theta\chi_\omega - P_1\rho_\theta) \tag{20}$$

$$w_\theta = -(\rho_\theta\chi_\omega - P_2\rho_\theta) \tag{21}$$

and

$$q_\theta = L\rho_\theta. \tag{22}$$

The pair $((v_\theta, w_\theta); q_\theta)$ is such that (3)-(5) hold. Moreover, we have

$$\begin{aligned} \|\rho_\theta\|_V &\le C\|\theta h\|_{L^2(Q)}, &(23)\\ \|v_\theta\|_{L^2(\omega\times(0,T))} &\le C\|\theta h\|_{L^2(Q)}, &(24)\\ \|w_\theta\|_{L^2(\omega\times(0,T))} &\le C\|\theta h\|_{L^2(Q)}, &(25)\\ \|q_\theta\|_{L^2(0,T,H^1(\Omega))} &\le C\|\theta h\|_{L^2(Q)}, &(26)\end{aligned}$$

where C is a positive constant dependant only for Ω, ω, a_0, T, Y_λ and $\mathcal{M}$.

Proof. Since $\rho_\theta \in V$ it follows that $v_\theta = -(\rho_\theta\chi_\omega - P_1\rho_\theta) \in L^2(\omega \times (0,T))$, $w_\theta = -(\rho_\theta\chi_\omega - P_2\rho_\theta) \in L^2(\omega \times (0,T))$ and $q_\theta \in L^2(0,T,H^1(\Omega))$. Since $P_1\rho_\theta \in Y_\lambda$ and $P_2\rho_\theta \in \mathcal{M}$ we have $v_\theta = -(\rho_\theta\chi_\omega - P_1\rho_\theta) \in Y_\lambda^\perp$ and $w_\theta = -(\rho_\theta\chi_\omega - P_2\rho_\theta) \in \mathcal{M}^\perp$. Substitute $-(\rho_\theta\chi_\omega - P_1\rho_\theta)$, $-(\rho_\theta\chi_\omega - P_2\rho_\theta)$ and $L\rho_\theta$ respectively by v_θ, w_θ and q_θ in the formula (19), it follows that

$$\begin{aligned} &\int_Q q_\theta L\rho dxdt - \int_0^T\int_\omega v_\theta(\rho - P_1\rho)dxdt \\ -\int_0^T\int_\omega w_\theta(\rho - P_2\rho)dxdt &= \int_Q h\rho dxdt, \;\; \forall \rho \in V.\end{aligned}$$

Taking into account that $P_1\rho \in Y_\lambda$ et $P_2\rho \in \mathcal{M}$, the above identity reduces to

$$\int\limits_Q q_\theta L\rho dxdt = \int\limits_Q h\rho dxdt + \int_0^T \int_\omega v_\theta \rho dxdt + \int_0^T \int_\omega w_\theta \rho dxdt \ \ \forall \rho \in V. \tag{27}$$

We show now that q_θ is in fact the weak solution by transposition of a backward heat problem. More precisely, if $f \in L^2(Q)$, let z be the solution of

$$\begin{cases} \dfrac{\partial z}{\partial t} - \Delta z + a_0 z &= f \quad in \quad Q \\ \dfrac{\partial z}{\partial \nu} + bz &= 0 \quad on \quad \Sigma \\ z(0) &= 0 \quad in \quad \Omega \end{cases} \tag{28}$$

Then $z \in V$ and so

$$\int\limits_Q q_\theta f dxdt = \int\limits_Q hzdxdt + \int_0^T \int_\omega v_\theta z dxdt \tag{29}$$

$$+ \int_0^T \int_\omega w_\theta z dxdt \ \ \forall z \in V.$$

Therefore q_θ is the weak solution by transposition of problem (4) with $v = v_\theta$ and $w = w_\theta$ (see J. L. Lions [16]). We know that the solution of this equation is in $L^2(0,T,H^1(\Omega)) \cap C\left((0,T), L^2(\Omega)\right)$ .Therefore $\dfrac{\partial q_\theta}{\partial \nu} \in L^2(0,T,H^{-\frac{1}{2}}(\Omega))$. Multiplying the first equation of (4) by $\rho \in \mathcal{V}$ and integrating by parts over Q, it following that

$$\begin{aligned} &- \int_\Omega q_\theta(T)\rho(T)dx + \int_\Omega q_\theta(0)\rho(0)dx + \int_Q q_\theta L\rho dxdt \\ &= \int\limits_Q h\rho dxdt + \int_0^T \int_\omega v_\theta \rho dxdt + \int_0^T \int_\omega w_\theta \rho dxdt \ \ \forall \rho \in \mathcal{V}. \end{aligned}$$

Since $\rho \in \mathcal{V}$ we deduce from (27) that

$$\int_\Omega q_\theta(0)\rho(0)dx = 0 \ \ \rho \in \mathcal{V}$$

There $q_\theta(0) = 0$ in Ω. Hence the first statement of proposition 5 is proved. It remains to prove the estimates (23)-(26). We set $\rho = \rho_\theta$ in (19). It follows from (19) that

$$\begin{aligned} a(\rho_\theta, \rho_\theta) &= ||q_\theta||^2_{L^2(0,T,H^1(\Omega))} + ||v_\theta||^2_{L^2(\omega\times(0,T))} + ||w_\theta||^2_{L^2(\omega\times(0,T))} \\ &\leq ||\theta h||_{L^2(Q)}||\rho_\theta||_\theta \\ &\leq C||\theta h||_{L^2(Q)}||\rho_\theta||_V. \end{aligned} \tag{30}$$

From (16) we obtain (23) and thus (24) and (25). Finally, (26) is a consequence of (24), (25) and classical properties of the heat equation. □

The adapted observability inequality (14) shows that the choice of the scalar product on $\mathcal{V}$ is not unique. Thus there exists infinitely many control functions (v_θ, w_θ) such that (3)-(5) hold.

Let consider the set of control variables

$$(v_\theta, w_\theta) \in L^2(\omega \times (0, T)) \times L^2(\omega \times (0, T))$$

such that (3)-(5) hold. By proposition 5 this set is nonempty and it is clearly convex and closed in $L^2(\omega \times (0, T)) \times L^2(\omega \times (0, T))$. Therefore, there exists a unique *pair* of variables *controls* (v_θ, w_θ) of minimal norm in $L^2(\omega \times (0, T))$ such that $(\widehat{v}_\theta, \widehat{w}_\theta, \widehat{q}_\theta = \widehat{q}_\theta(x, t; (\widehat{v}_\theta, \widehat{w}_\theta)))$ be the solution of (3)-(6).

3. Optimality System for the Optimal Solution

3.1. Penalization

The optimal solution $(\widehat{v}_\theta, w_\theta, \widehat{q}_\theta)$ can be approximated considering the penalization method by J. L. Lions [17]. Let $\varepsilon > 0$. Define the functional

$$\begin{aligned} J_\varepsilon(v, w, q) &= \frac{1}{2}\|v\|^2_{L^2(\omega\times(0,T))} + \frac{1}{2}\|w\|^2_{L^2(\omega\times(0,T))} \\ &\quad + \frac{1}{2\varepsilon}\| - \frac{\partial q}{\partial t} - \Delta q + a_0 q - h - v\chi_\omega - w\chi_\omega\|^2_{L^2(Q)}. \end{aligned} \tag{31}$$

for any (v, w, q) such that

$$\begin{cases} v \in Y_\lambda^\perp, \ w \in \mathcal{M}^\perp \\ -\dfrac{\partial q}{\partial t} - \Delta q + a_0 q \in L^2(Q) \\ \dfrac{\partial q}{\partial \nu} + bq = 0 \ on \ \Sigma, \ q(0) = q(T) = 0 \text{ in } \Omega \end{cases} \tag{32}$$

Consider the minimization problem

$$\min J_\varepsilon(v, w, q), (v, w, q) \text{ subject to } (32) \tag{33}$$

Proposition 6. *Under the assumptions of proposition 5, the minimization problem (33) has an optimal solution. In other words, there exists* $(v_\varepsilon, w_\varepsilon, q_\varepsilon)$ *such that*

$$J_\varepsilon(v_\varepsilon, w_\varepsilon, q_\varepsilon) = \min\{J_\varepsilon(v, w, q)|(v, w, q) \textit{ subject to } 32\} \tag{34}$$

Proof. Since the triplet $(v_\theta, w_\theta, q_\theta)$ define in the Proposition 5 satisfying (32) and $J_\varepsilon(v, w, q) \geq 0$, we can define a real d_ε such that

$$d_\varepsilon = \min\{J_\varepsilon(v, w, q), (v, w, q) \text{ subject to } (32)\}. \tag{35}$$

Let (v_n, w_n, q_n) be a minimizing sequence satisfying (32) and such that

$$d_\varepsilon \leq J_\varepsilon(v_n, w_n, q_n) < d_\varepsilon + \frac{1}{n} < d_\varepsilon + 1.$$

In particular,

$$0 < d_\varepsilon \leq J_\varepsilon(v_\theta, w_\theta, q_\theta) = \frac{1}{2}||v_\theta||^2_{L^2(\omega\times(0,T))} + \frac{1}{2}||w_\theta||^2_{L^2(\omega\times(0,T))}.$$

It follows from the estimates (24) and (25) that there exists a constante C, independant of n such that

$$J_\varepsilon(v_n, w_n, q_n) \leq C.$$

Therefore, from the form of J_ε given by (31), we obtain

$$||v_n||_{L^2(\omega\times(0,T))} \leq C, \tag{36}$$

$$||w_n||_{L^2(\omega\times(0,T))} \leq C \tag{37}$$

$$|| -\frac{\partial q_n}{\partial t} - \Delta q_n + a_0 q_n - h - v_n\chi_\omega + w_n\chi_\omega||_{L^2(Q)} \leq C\sqrt{\varepsilon}. \tag{38}$$

From (36) and (37), there exist v_ε and w_ε in $L^2(\omega \times (0,T))$ and the subsequences of (v_n) and (w_n) (still denoted by (v_n) and (w_n)) such that

$$\begin{cases} v_n \rightharpoonup v_\varepsilon & L^2(\omega \times (0,T)), \\ w_n \rightharpoonup w_\varepsilon & L^2(\omega \times (0,T)). \end{cases} \tag{39}$$

Since $v_n \in Y_\lambda^\perp$ and $w_n \in \mathcal{M}^\perp$ with $Y_\lambda^\perp$ and $\mathcal{M}^\perp$ some vectorial closed subspaces of $L^2(\omega \times (0,T))$, we have

$$v_\varepsilon \in Y_\lambda^\perp \text{ and } w_\varepsilon \in \mathcal{M}^\perp. \tag{40}$$

Since a consequence of (32), (36), (37), (38) we have,

$$\|q_n\|_{L^2(0,T,H^1(\Omega))} \leq C.$$

Thus, there exists $q_\varepsilon \in L^2(0,T,H^1(\Omega))$ and a subsequence (q_n) (still denoted (q_n)) such that

$$q_n \rightharpoonup q_\varepsilon \quad \text{weakly in} \quad L^2(0,T,H^1(\Omega)). \tag{41}$$

Moreover, q_ε satisfies

$$\begin{cases} -\dfrac{\partial q_\varepsilon}{\partial t} - \Delta q_\varepsilon + a_0 q_\varepsilon \in L^2(Q), \\ \dfrac{\partial q_\varepsilon}{\partial \nu} + b q_\varepsilon = 0 \text{ on } \Sigma, \quad q_\varepsilon(T) = q_\varepsilon(0) = 0 \ in \ \Omega. \end{cases} \tag{42}$$

Finally, from (40) and (42), we deduce that $(v_\varepsilon, w_\varepsilon, q_\varepsilon)$ satisfies (32). Combining (35), (39), (41) and the lower semicontinuity of the functional J_ε, we obtain that $J_\varepsilon(v_\varepsilon, w_\varepsilon, q_\varepsilon) \leq \liminf J_\varepsilon(v_n, w_n, q_n) = d_\varepsilon$. In other words $(v_\varepsilon, w_\varepsilon, q_\varepsilon)$ is the optimal solution of problem (34). The uniqness of $(v_\varepsilon, w_\varepsilon, q_\varepsilon)$ is the consequence of the strict convexity of J_ε. □

Proposition 7. *Under the assumptions of proposition 5, the triplet* $(v_\varepsilon, w_\varepsilon, q_\varepsilon)$ *is the optimal solution (34) if and only if there exists a function* ρ_ε *such that triplet* $((v_\varepsilon, w_\varepsilon), q_\varepsilon, \rho_\varepsilon)$ *satisfies the following approximate optimality condition:*

$$\begin{cases} -\dfrac{\partial q_\varepsilon}{\partial t} - \Delta q_\varepsilon + a_0 q_\varepsilon & = & h\chi_O + v_\varepsilon \chi_\omega + w_\varepsilon \chi_\omega + \varepsilon \rho_\varepsilon \ in \ Q \\ \dfrac{\partial q_\varepsilon}{\partial \nu} + b q_\varepsilon & = & 0 \qquad on \quad \Sigma \\ q_\varepsilon(T) & = & 0 \qquad in \quad \Omega \end{cases} \tag{43}$$

$$q_\varepsilon(0) = 0 \ \ in \ \ \Omega \tag{44}$$

$$\left\{ \begin{array}{rcl} \dfrac{\partial \rho_\varepsilon}{\partial t} - \Delta \rho_\varepsilon + a_0 \rho_\varepsilon & = & 0 \ \ in \ \ Q \\ \dfrac{\partial \rho_\varepsilon}{\partial \nu} + b \rho_\varepsilon & = & 0 \ \ on \ \ \Sigma \end{array} \right. \tag{45}$$

$$v_\varepsilon \ \ = \ \ -(\rho_\varepsilon \chi_\omega - P_1 \rho_\varepsilon) \ \ in \, \omega \times (0, T), \tag{46}$$

$$w_\varepsilon \ \ = \ \ -(\rho_\varepsilon \chi_\omega - P_2 \rho_\varepsilon) \ \ in \, \omega \times (0, T). \tag{47}$$

Proof. Express the Euler-Lagrange optimality conditions which characterize $(v_\varepsilon, w_\varepsilon, q_\varepsilon)$:

$$\frac{d}{d\lambda} J_\varepsilon(v_\varepsilon, w_\varepsilon, q_\varepsilon + \lambda \varphi)|_{\lambda=0} = 0$$

for any $\varphi \in C^\infty(\overline{Q})$ such that $\dfrac{\partial \varphi}{\partial \nu} + b\varphi = 0$ on Σ, $\varphi(0) = \varphi(T) = 0$ in Ω,

$$\frac{d}{d\lambda} J_\varepsilon(v_\varepsilon + \lambda v, w_\varepsilon, q_\varepsilon)|_{\lambda=0} = 0 \ \ \forall \, v \in Y_\lambda^\perp \ ,$$

$$\frac{d}{d\lambda} J_\varepsilon(v_\varepsilon, w_\varepsilon + \lambda w, q_\varepsilon)|_{\lambda=0} = 0 \ \ \forall w \in \mathcal{M}^\perp.$$

After some calculations, we have

$$\int_Q \frac{1}{\varepsilon}(-\frac{\partial q_\varepsilon}{\partial t} - \Delta q_\varepsilon + a_0 q_\varepsilon - h_0 - v_\varepsilon \chi_\omega - w_\varepsilon \chi_\omega)(-\frac{\partial \varphi}{\partial t} - \Delta \varphi + a_0 \varphi) dx dt = 0 \tag{48}$$

for any $\varphi \in C^\infty(\overline{Q})$ such that $\dfrac{\partial \varphi}{\partial \nu} + b\varphi = 0$ on Σ, $\varphi(0) = \varphi(T) = 0$ in Ω,

$$\int_0^T \int_\omega v_\varepsilon v dx dt - \int_Q \frac{1}{\varepsilon}(-\frac{\partial q_\varepsilon}{\partial t} - \Delta q_\varepsilon + a_0 q_\varepsilon - h_0 - v_\varepsilon \chi_\omega - w_\varepsilon \chi_\omega) v dx dt = 0 \tag{49}$$

$\forall v \in Y_\lambda^\perp$ and

$$\int_0^T \int_\omega w_\varepsilon w dx dt - \int_Q \frac{1}{\varepsilon}(-\frac{\partial q_\varepsilon}{\partial t} - \Delta q_\varepsilon + a_0 q_\varepsilon - h_0 - v_\varepsilon \chi_\omega - w_\varepsilon \chi_\omega) w dx dt = 0 \tag{50}$$

$\forall w \in \mathcal{M}^\perp$.

We set

$$\rho_\varepsilon = -\frac{1}{\varepsilon}(-\frac{\partial q_\varepsilon}{\partial t} - \Delta q_\varepsilon + a_0 q_\varepsilon - h_0 - v_\varepsilon \chi_\omega - w_\varepsilon \chi_\omega). \tag{51}$$

Then (48), (49) and (50) became respectively

$$\int_Q \rho_\varepsilon(-\frac{\partial \varphi}{\partial t} - \Delta\varphi + a_0\varphi) dx dt = 0 \tag{52}$$

for any $\varphi \in C^\infty(\overline{Q})$ such that $\frac{\partial \varphi}{\partial \nu} + b\varphi = 0$ on Σ, $\varphi(0) = \varphi(T) = 0$ in Ω,

$$\int_0^T \int_\omega v_\varepsilon v dx dt + \int_Q \rho_\varepsilon v dx dt = 0, \ \forall v \in Y_\lambda^\perp \tag{53}$$

and

$$\int_0^T \int_\omega w_\varepsilon w dx dt + \int_Q \rho_\varepsilon w dx dt = 0, \ \ \forall w \in \mathcal{M}^\perp. \tag{54}$$

Consider the first part of (52), we deduce that

$$\frac{\partial \rho_\varepsilon}{\partial t} - \Delta \rho_\varepsilon + a_0 \rho_\varepsilon = 0 \text{ in } Q$$

so $\rho_\varepsilon \in L^2(Q)$ with $L\rho_\varepsilon \in L^2(Q)$. Then we can define on the one hand ρ_ε and $\frac{\partial \rho_\varepsilon}{\partial \nu}$ on Γ and on the other hand we prove that $\frac{\partial \rho_\varepsilon}{\partial \nu} + b\, \rho_\varepsilon = 0$ on Σ.

Now, we consider (53)

$$\int_0^T \int_\omega (v_\varepsilon + \rho_\varepsilon) v dx dt = 0 \ \ \forall v \in Y_\lambda^\perp.$$

Thus $v_\varepsilon + \rho_\varepsilon \chi_\omega \in Y_\lambda$. Since $v_\varepsilon \in Y_\lambda^\perp$, we have $v_\varepsilon + \rho_\varepsilon \chi_\omega = P_1(v_\varepsilon + \rho_\varepsilon \chi_\omega) = P_1 \rho_\varepsilon$ and therefore $v_\varepsilon = -(\rho_\varepsilon \chi_\omega - P_1 \rho_\varepsilon)$.

Finally we consider (54)

$$\int_0^T \int_\omega (w_\varepsilon + \rho_\varepsilon) w dx dt = 0 \; \forall w \in \mathcal{M}^\perp$$

we have therefore $v_\varepsilon + \rho_\varepsilon \chi_\omega \in \mathcal{M}^\perp$. Since $w_\varepsilon \in \mathcal{M}^\perp$, we have $w_\varepsilon + \rho_\varepsilon \chi_\omega = P_2(w_\varepsilon + \rho_\varepsilon \chi_\omega) = P_2\rho_\varepsilon$ and thus $w_\varepsilon = -(\rho_\varepsilon \chi_\omega - P_2\rho_\varepsilon)$. The proposition is proved. □

Remark 8. *There is no information avaible concerning $\rho_\varepsilon(0)$ and $\rho_\varepsilon(T)$.*

Proposition 9. *Let v_ε, w_ε, q_ε and ρ_ε be the functions defined in the proposition 7. Then for $\varepsilon \to 0$, we have the follow limits:*

$$\begin{aligned} v_\varepsilon &\rightharpoonup \widehat{v}_\theta \textit{ weakly in } L^2(\omega \times (0,T)), & (55)\\ w_\varepsilon &\rightharpoonup \widehat{w}_\theta \textit{ weakly in } L^2(\omega \times (0,T)) & (56)\\ q_\varepsilon &\rightharpoonup \widehat{q}_\theta \textit{ weakly in } L^2(0,T,H^1(\Omega)), & (57)\\ \rho_\varepsilon &\rightharpoonup \widehat{\rho}_\theta \textit{ weakly in } V. & (58)\end{aligned}$$

Proof. We show the proposition in four steps.

Step 1. We now look for a priori estimates for the approximate controls and state v_ε, w_ε, q_ε and ρ_ε.

From (38), (36) and (37) we have

$$\begin{aligned} \left|\left| -\frac{\partial q_\varepsilon}{\partial t} - \Delta q_\varepsilon + a_0 q_\varepsilon - h - v_\varepsilon \chi_\omega - w_\varepsilon \chi_\omega \right|\right|_{L^2(Q)} &\leq C\sqrt{\varepsilon}, & (59)\\ ||v_\varepsilon||_{L^2(\omega\times(0,T))} &\leq C, & (60)\\ ||w_\varepsilon||_{L^2(\omega\times(0,T))} &\leq C. & (61)\end{aligned}$$

and in view of (51) and (59)

$$||_\varepsilon \rho_\varepsilon||_{L^2(Q)} \leq C\sqrt{\varepsilon}$$

Since q_ε satisfies (32), we obtain from (59), (60) and (61) the following estimate

$$||q_\varepsilon||_{W(0,T)} \leq C\left(1+\sqrt{\varepsilon}\right). \tag{62}$$

where $W(0,T) = \left\{\varphi / \varphi \in L^2(0,T,H^1(\Omega)) \text{ and } \frac{\partial \varphi}{\partial t} \in L^2(0,T,H^{-1}(\Omega))\right\}$

From (46) and (60), we obtain

$$||-\rho_\varepsilon + P_1\rho_\varepsilon||_{L^2(\omega\times(0,T))} \leq C. \tag{63}$$

In the same way, from (47) and (61), we have

$$||-\rho_\varepsilon + P_2\rho_\varepsilon||_{L^2(\omega\times(0,T))} \leq C. \tag{64}$$

Since $L\rho_\varepsilon = 0$, using the definition of the norm on V given by (16), we have

$$||\rho_\varepsilon||_V \leq C. \tag{65}$$

Since $\rho_\varepsilon \in \mathcal{V}$, applying the observability inequality (14) to ρ_ε, we have $\left\|\frac{1}{\theta}\rho_\varepsilon\right\|_{L^2(\omega\times(0,T))} \leq C$. Then, using (63), (64) and the fact that $\frac{1}{\theta}$ is in $L^\infty(Q)$, we deduce that

$$\left\|\frac{1}{\theta}P_1\rho_\varepsilon\right\|_{L^2(\omega\times(0,T))} \leq C \text{ and } \left\|\frac{1}{\theta}P_2\rho_\varepsilon\right\|_{L^2(\omega\times(0,T))} \leq C.$$

Since $P_1\rho_\varepsilon \in Y_\lambda$ and $P_2\rho_\varepsilon \in \mathcal{M}$ with Y_λ and $\mathcal{M}$ finite dimensional, we have $||P_1\rho_\varepsilon||_{L^2(\omega\times(0,T))} \leq C$, $||P_2\rho_\varepsilon||_{L^2(\omega\times(0,T))} \leq C$. Thus using again (63) et (64), we obtain

$$||\rho_\varepsilon||_{L^2(\omega\times(0,T))} \leq C, \tag{66}$$

Step 2. We study the convergence of $(v_\varepsilon, w_\varepsilon, q_\varepsilon)$.
From (60), (61) and (62) we can extract some subsequences still denoted $(q_\varepsilon)_\varepsilon$, $(v_\varepsilon)_\varepsilon$ and $(w_\varepsilon)_\varepsilon$ such that

$$v_\varepsilon \rightharpoonup v_0 \text{ weakly in } L^2(\omega\times(0,T)), \tag{67}$$
$$w_\varepsilon \rightharpoonup w_0 \text{ weakly in } L^2(\omega\times(0,T)), \tag{68}$$
$$q_\varepsilon \rightharpoonup q_0 \text{ weakly in } W(0,T). \tag{69}$$

Since $v_\varepsilon \in Y_\lambda^\perp$ and $w_\varepsilon \in \mathcal{M}^\perp$ with Y_λ and $\mathcal{M}$ the real closed subspaces of $L^2(\omega\times(0,T))$, we have

$$v_0 \in Y_\lambda^\perp \text{ and } w_0 \in \mathcal{M}^\perp. \tag{70}$$

Since the injection from $W(0,T)$ into $L^2(Q)$ is compact, (see [19] P 57, [18]) the pair $((v_0, w_0), q_0)$ is such that

$$\left\{\begin{array}{rcll} -\dfrac{\partial q_0}{\partial t} - \Delta q_0 + a_0 q_0 & = & h_0 + v_0\chi_\omega + w_0\chi_\omega & \text{in} \quad Q, \\ \dfrac{\partial q_0}{\partial \nu} + b q_0 & = & 0 & \text{on} \quad \Sigma, \\ q_0(T) & = & 0 & \text{in} \quad \Omega, \end{array}\right. \tag{71}$$

$$q_0(0) = 0 \text{ in } \Omega.$$

Step 3. We show that $(v_0, w_0, q_0 = q_0(x,t;(v_0,w_0))) = (\widehat{v}_\theta, w_\theta, \widehat{q}_\theta = \widehat{q}_\theta(x,t;(\widehat{v}_\theta, \widehat{w}_\theta)))$.

From the expression of J_ε given by (31), we can write

$$\frac{1}{2}\,||v_\varepsilon||^2_{L^2(\omega\times(0,T))} + \frac{1}{2}\,||w_\varepsilon||^2_{L^2(\omega\times(0,T))} \leq J_\varepsilon(v_\varepsilon, q_\varepsilon).$$

Since $(\widehat{v}_\theta, \widehat{w}_\theta, \widehat{q}_\theta)$ satisfies (3)-(5), we have

$$\begin{aligned} &\frac{1}{2}\,||v_\varepsilon||^2_{L^2(\omega\times(0,T))} + \frac{1}{2}\,||w_\varepsilon||^2_{L^2(\omega\times(0,T))} \leq J_\varepsilon(v_\varepsilon, q_\varepsilon) \\ &\leq \frac{1}{2}\,\|\widehat{v}_\theta\|^2_{L^2(\omega\times(0,T))} + \frac{1}{2}\,\|\widehat{w}_\theta\|^2_{L^2(\omega\times(0,T))}. \end{aligned} \tag{72}$$

Thus using (67) and (68) and taking a limit in (72), we obtain

$$\begin{aligned} &\frac{1}{2}\,||v_0||^2_{L^2(\omega\times(0,T))} + \frac{1}{2}\,||w_0||^2_{L^2(\omega\times(0,T))} \leq \liminf_{\varepsilon\to 0} J_\varepsilon(v_\varepsilon, q_\varepsilon) \\ &\leq \frac{1}{2}\,\|\widehat{v}_\theta\|^2_{L^2(\omega\times(0,T))} + \frac{1}{2}\,\|\widehat{w}_\theta\|^2_{L^2(\omega\times(0,T))}, \end{aligned}$$

consequently,

$$||v_0||^2_{L^2(\omega\times(0,T))} + ||w_0||^2_{L^2(\omega\times(0,T))} \leq \|\widehat{v}_\theta\|^2_{L^2(\omega\times(0,T))} + \|\widehat{w}_\theta\|^2_{L^2(\omega\times(0,T))}$$

since the triplet $(\widehat{v}_\theta, w_\theta, \widehat{q}_\theta)$ is the optimal solution we have

$$\begin{aligned} ||v_0||^2_{L^2(\omega\times(0,T))} + ||w_0||^2_{L^2(\omega\times(0,T))} &= \|\widehat{v}_\theta\|^2_{L^2(\omega\times(0,T))} + \|\widehat{w}_\theta\|^2_{L^2(\omega\times(0,T))}, \\ ||(v_0, w_0)||^2_{L^2(\omega\times(0,T))} &= ||(\widehat{v}_\theta, \widehat{w}_\theta)||^2_{L^2(\omega\times(0,T))}, \\ ||(v_0, w_0)||_{L^2(\omega\times(0,T))} &= ||(\widehat{v}_\theta, \widehat{w}_\theta)||_{L^2(\omega\times(0,T))}. \end{aligned}$$

From the unicity of $\widehat{v}_\theta$ and $\widehat{w}_\theta$ we have $v_0 = \widehat{v}_\theta$ and $w_0 = \widehat{w}_\theta$.

Step 4. From (66) et (65), there exists a subsequence, still denoted $(\rho_\varepsilon)_\varepsilon$ and $\widehat{\rho}_\theta \in V$ such that

$$\begin{aligned} \rho_\varepsilon &\rightharpoonup \widehat{\rho}_\theta \text{ weakly in } V & (73)\\ \rho_\varepsilon &\rightharpoonup \widehat{\rho}_\theta \text{ weakly in } L^2(\omega \times (0, T)). & (74)\end{aligned}$$

Since P is the compact operator, we deduce from (74) that

$$\begin{aligned} P_1\rho_\varepsilon &\rightarrow P_1\widehat{\rho}_\theta \text{ weakly in } L^2(\omega \times (0, T)), & (75)\\ P_2\rho_\varepsilon &\rightarrow P_2\widehat{\rho}_\theta \text{ weakly in } L^2(\omega \times (0, T)). \end{aligned}$$

Combuning then (74) and (75), we obtain

$$\begin{aligned} v_\varepsilon &= \rho_\varepsilon\chi_\omega - P\rho_\varepsilon \rightharpoonup \widehat{v}_\theta = \widehat{\rho}_\theta\chi_\omega - P_1\widehat{\rho}_\theta \text{ weakly in } L^2(\omega \times (0, T)),\\ w_\varepsilon &= \rho_\varepsilon\chi_\omega - P_2\rho_\varepsilon \rightharpoonup \widehat{w}_\theta = \widehat{\rho}_\theta\chi_\omega - P_2\widehat{\rho}_\theta \text{ weakly in } L^2(\omega \times (0, T)). \end{aligned}$$

This achieve the proof of existence. □

4. Discriminating Sentinels with Given Sensivity

In this section, we apply the precedent results to obtain information on the pollution parameters in a problem gouverned by the semi linear parabolic equation with Fourier boundary condition.

More precisely, let N, $M_1 \in \mathbb{N}\backslash\{0\}$ and let Ω be a bounded subset of $\mathbb{R}^N$ with boundary Γ of class C^2. For any $T > 0$, we set $Q = \Omega \times (0, T)$ and $\Sigma = \partial\Omega \times (0, T) = \Gamma \times (0, T)$. We consider now the system modelling a follow pollution problem:

$$\begin{cases} \dfrac{\partial y}{\partial t} - \Delta y + f(y) &= \xi + \dfrac{\partial y}{\partial \nu} + by \\ \qquad = & 0 \text{ on } \Sigma \\ \quad y(0) &= y^0 + \tau\,\hat{y}^0 \text{ in } \Omega \end{cases} \quad (76)$$

Model (76) can be used to describe a pollution in a lake or a river. In this case:

- $y : Q \longrightarrow \mathbb{R}$ is a unknow function which represents for example the polluant concentration;

- f is a real-valued given function of class C^1, globally lipschitz;
- The source term is unknown and represents pollution source of the form $\xi + \sum_{i=1}^{M_1} \lambda_i \hat{\xi}_i$. The functions ξ and $\{\hat{\xi}_i\}_{i=1}^{M_1}$ are known whereas the real coefficients $\{\lambda_i\}_{i=1}^{M_1}$ are unknown;
- The initial condition is of the form $y^0 + \tau\, \hat{y}^0$ where the function y^0 is known while τ real, is unknown.

  We assume that:

- $b \in L^\infty(\Sigma)$, $b \geq 0$,
- y^0 and $\hat{y}^0$ belong to $L^2(\Omega)$, ξ and ξ_i belong to $L^2(Q)$,
- the functions ξ_i, $1 \leq i \leq M_1$ are linearly independent,
- the real τ is sufficiently small,
- the function f verifies

$$f(0) = 0 \tag{77}$$

and satisfies the growth condition:

$$\begin{aligned} &|f(s_1) - f(s_2) - f'(0)(s_1 - s_2)\,| \\ &\leq c(\,|s_1|^{P-1} + |s_2|^{P-1})|s_1 - s_2| \;\; \forall s_1, s_2 \in \mathbb{R} \end{aligned} \tag{78}$$

for some $c > 0$ and $p > 1$ such that $p < \dfrac{N+4}{N}$.

We assume without loss the generality of the problem that

$$\xi = 0 \;\text{ in } Q \;\text{ and } y^0 = 0 \text{ in } \Omega. \tag{79}$$

Under the above conditions on the data, it proved in [22], [12] that there exists $\alpha > 0$ such that if

$$||\tau \hat{y}^0||_{L^2(\Omega)} + ||\sum_{i=1}^{M_1} \lambda_i \hat{\xi}_i\,||_{L^2(Q)} \leq \alpha$$

The problem (76) admits a unique solution in $C((0,T);\ L^2(\Omega)) \cap L^2(0,T;H^1(\Omega))$. Moreover, if we denote by I a neighbourhood of 0, the applications

$$\begin{aligned} \tau \quad \rightarrow \quad & y(\lambda_i,\tau) \text{ and } \lambda_i \rightarrow y(\lambda_i,\tau),\ (1 \leq i \leq M_1) \\ & \text{are in } \mathcal{C}\left(I, L^2(0,T;L^2(\Omega))\right). \end{aligned} \tag{80}$$

for simplify, we denote

$$y(x,t;\lambda,\tau) = y(\lambda,\tau).$$

the unique solution of (76).

We want to obtain by observation of (76) some informations on the pollution term $\sum_{i=1}^{M_1} \lambda_i \hat{\xi}_i$ without calculing the missing term $\tau\hat{y}^0$.

In the model (76), we are interested in identifying the parameters λ_i without any attempt of computing the missing term $\tau\hat{y}^0$. To identify these parameters, we use the method of sentinels due to J.L.Lions dans [15]. It was revisited by O. Nakoulima in [20].

The theory of sentinels introduced by J.L. Lions relies on three considerations:

1. A state equation represented here by (76) whose solution $y = y(x,t,\lambda,\tau) = y(\lambda,\tau)$ depends on two families of parameters $\lambda = \{\lambda_1, \ldots.\lambda_{M_1}\}$ and τ.

2. An observation

$$y_{obs} = m_0 + \sum_{i=1}^{M_2} \beta_i m_i, \tag{81}$$

   where $M_2 \in \mathbb{N}$; $m_0, m_1, m_2, \cdots, m_{M_2} \in L^2(O \times (O,T))$ are given by measurement of y in $L^2(O \times (O,T))$. The real-valued β_i are unknow and we suppose that they are small enough. The terms $\beta_i m_i$ are the interference terms. m_0 which is a measurement of y if there is no noise. The functions m_i are lineary independant on $O \times (0,T)$. O a non-empty open subset of Ω called observatory.

3. A function $S = S(\lambda, \tau)$ called "sentinel" defined for $h \in L^2(O \times (0,T))$ and a non-empty open subset ω of O ($\omega \subset \overline{\omega} \subset O$) by

$$S(\lambda,\tau)(u) = \int_0^T \int_O hy(\lambda,\tau)dxdt + \int_0^T \int_\omega uy(\lambda,\tau)dxdt \qquad (82)$$

where the control function u is to be found of minimal norm in $L^2(\omega \times (0,T))$ among functions S defined in (82) and satisfying the following conditions:

- S is stationary to the first order with respect to the missing term $\tau\, \hat{y}^0$, i.e.,

$$\frac{\partial S}{\partial \tau}(0,0) = 0 \ \forall \hat{y}^0. \qquad (83)$$

- S is stationary to the first order with respect to the interference terms $\beta_i m_i$, i.e.,

$$\int_0^T \int_O h_0 m_i dxdt + \int_0^T \int_\omega u m_i dxdt = 0, \ 1 \leq i \leq M_2. \qquad (84)$$

- S is sensitive to the first order with respect to the pollution terms $\lambda_i \hat{\xi}_i$:

$$\frac{\partial S}{\partial \lambda_i}(0,0) = c_i, 1 \leq i \leq M_1 \qquad (85)$$

where c_i, $(1 \leq i \leq M_1)$, are some given constants not all identicaly null.

- We look for the control u in the form:

$$\begin{aligned} u &= u_0 + \widehat{v} + \widehat{w} \qquad (86) \\ &\text{with} \\ u_0 &\in Y_\lambda + \mathcal{M}, \widehat{v} \in Y_\lambda^\perp \text{ and } \widehat{w} \in \mathcal{M}^\perp. \end{aligned}$$

Where Y_λ and $\mathcal{M}$ are them defined in (2). We choose then u among the controls called "admissibles", i.e., the controls for which the function S defined in (82) satisfying (83)-(85), with,

$$||(\widehat{v}, \widehat{w})|| = \min_{(v,w)\in E} \{||(v,w)||\} \qquad (87)$$

where

$$E = \{(v,w) \in Y_\lambda \times \mathcal{M}, \text{ such that } (u, S(u)) \text{ satisfies } (83) - (85)\}.$$

Using the adjoint problem, one shows that the existence of these sentinels is reduced to the solution of null-controllability problem (we can see [3], [4]). In this chapter we propose to study a null controlability problem for Fourier boundary condition using a discriminating sentinel with given sensitivity.

Definition 10. *We will refer to the function S given by (82)-(85) as a discriminating sentinel with given $\{c_i\}$ sensitivity.*

Let y_0 be the solution of (76) when $\lambda = 0$ and $\tau = 0$. Then, in view of (79), we have

$$y_0 = 0 \text{ in } Q \tag{88}$$

$y_\tau = \dfrac{d}{d\tau} y(\lambda,\tau)|_{\tau=0}$ and $y_{\lambda_i} = \dfrac{d}{d\lambda_i} y(\lambda,\tau)|_{\lambda_i=0}$ are respectively solution of

$$\begin{cases} \dfrac{\partial y_\tau}{\partial t} - \Delta y_\tau + f'(0) y_\tau &= 0 \text{ in } Q, \\ \dfrac{\partial y_\tau}{\partial \nu} + b y_\tau &= 0 \text{ on } \Sigma, \\ y_\tau(0) &= \hat{y}^0 \text{ in } \Omega \end{cases} \tag{89}$$

and

$$\begin{cases} \dfrac{\partial y_{\lambda_i}}{\partial t} - \Delta y_{\lambda_i} + f'(0) y_{\lambda_i} &= \hat{\xi}_i \text{ in } Q, \\ \dfrac{\partial y_{\lambda_i}}{\partial \nu} + b y_{\lambda_i} &= 0 \text{ on } \Sigma, \\ y_{\lambda_i}(0) &= 0 \text{ in } \Omega \end{cases} \tag{90}$$

where $f'(0)$ denotes the derivative of f at $y_0 = 0$. Observing on the one hand that $f'(0) \in L^\infty(Q)$ and from the condition (78), and the other hand $\hat{y}^0 \in L^2(\Omega)$ and $b \in L^\infty(\Sigma)$ we deduce that the problems (89) and (90) admit respectively some unique solution $y_\tau \in C((0,T); L^2(\Omega)) \cap L^2(0,T; H^1(\Omega))$ et $y_{\lambda_i} \in L^2(0,T; H^2(\Omega)) \cap H^1(0,T; L^2(\Omega))$.

Let χ_ω be the characteristic function of the set ω. We set

$$Y_\lambda = vect(y_{\lambda_1}\chi_\omega, \cdots, y_{\lambda_{M_1}}\chi_\omega) \tag{91}$$

$$\mathcal{M} = vect(m_1\chi_\omega, \cdots, m_{M_2}\chi_\omega) \tag{92}$$

the vector subspace of $L^2(\omega \times (0,T))$ generated by the M_1 independent functions $y_{\lambda_i}\chi_\omega$ $(1 \le i \le M_1)$ and the M_2 independent functions $m_i\chi_\omega$ $(1 \le i \le M_2)$.

We also set

$$a_0 = f'(0) \tag{93}$$

4.1. Equivalence to the Null-Controlability Problem

Since y_τ and y_{λ_i} are respectively solutions of (89) and (90), the stationary condition (83)-(84) and respectively the sensitivity conditions (85) hold if and only if

$$\int_0^T \int_O h_0 y_\tau dxdt + \int_0^T \int_\omega u y_\tau dxdt = 0 \;\; \hat{y}^0 \in L^2(\Omega), \tag{94}$$

$$\int_0^T \int_O h m_i dxdt + \int_0^T \int_\omega u m_i dxdt = 0, \; 1 \le i \le M_2, \tag{95}$$

and

$$\int_0^T \int_O h_0 y_{\lambda_i} dxdt + \int_0^T \int_\omega u y_{\lambda_i} dxdt = c_i, \; 1 \le i \le M_1. \tag{96}$$

In order to transform equation (94) we introduce the classical adjoint state.

More precisely, we consider the solution $q = q(x, t, u)$ of the linear problem

$$\begin{cases} -\dfrac{\partial q}{\partial t} - \Delta q + a_0 q &= h\chi_O + u\chi_\omega \quad \text{in } Q \\ \dfrac{\partial q}{\partial \nu} + bq &= 0 \quad \text{on } \Sigma \\ q(T) &= 0 \quad \text{in } \Omega \end{cases} \tag{97}$$

where χ_O is characteristic function of the open set O. It well know that the problem (97) admits a unique solution q in $L^2(0, T, H^1(\Omega)) \cap C\left((0, T), L^2(\Omega)\right)$ and $\dfrac{\partial q}{\partial t} \in L^2(0, T, H^{-1}(\Omega))$(see [22]).

First, multiplying both sides of the differential equation in (97) by $y_\tau \in C([0, T], L^2(\Omega)) \cap L^2(0, T; H^1(\Omega))$ which is solution of (89), and integrating by parts in Q, we get

$$\int_0^T \int_O h_0 y_\tau dxdt + \int_0^T \int_\omega u y_\tau dxdt = \int_\Omega q(0)\widehat{y}^0 dx \;\; \forall \hat{y}^0 \in L^2(\Omega).$$

Thus, the condition (83) or (94) holds if and only if

$$q(0) = 0 \;\; in\; \Omega. \tag{98}$$

Then, multiplying both sides of the differential equation in (97) by $y_{\lambda_i} \in L^2(0,T;H^2(\Omega)) \cap H^1(0,T;L^2(\Omega))$ which is solution of (90), and integrating by parts in Q, we have

$$\int_0^T \int_\Omega q\widehat{\xi}_i dxdt = \int_0^T \int_O h_0 y_{\lambda_i} dxdt + \int_0^T \int_\omega u y_{\lambda_i} dxdt \;\; 1 \le i \le M_1.$$

Thus, the condition (85) or (96) is equivalent to

$$\int_0^T \int_\Omega q\widehat{\xi}_i dxdt = c_i \;\; 1 \le i \le M_1. \tag{99}$$

Let Y_λ be a *real* closed vector subspace defined in (91). Since Y_λ is a vector subspace $L^2(\omega\times(0,T))$ generated by the independant functions, and Y_λ is finite dimensional, there exists a unique $u_1 \in Y_\lambda$ such that

$$c_i - \int_0^T \int_O h_0 y_{\lambda_i} dxdt = \int_0^T \int_\omega u_1 y_{\lambda_i} dxdt \;\; 1 \le i \le M_2.$$

Therefore, the condition (96) or (99) holds if and only if

$$u - u_1 \in Y_\lambda^\perp.$$

there exists therefore $k_1 \in Y_\lambda^\perp$ such that

$$u = u_1 + k_1. \tag{100}$$

We consider now (95). Let $\mathcal{M}$ be a *real* closed vector subspace defined in (92). Since $\mathcal{M}$ is a vector subspace $L^2(\omega\times(0,T))$ generated by the independant functions, and $\mathcal{M}$ is finite dimensional, there exists a unique $u_2 \in \mathcal{M}$ such that

$$\int_0^T \int_O h_0 m_i dxdt = -\int_0^T \int_\omega u_2 m_i dxdt, \;\; 1 \le i \le M_1.$$

Therefore, the condition (95) holds if and only if

$$u - u_2 \in \mathcal{M}^\perp$$

there exists therefore $k_2 \in \mathcal{M}^\perp$ such that

$$u = u_2 + k_2. \tag{101}$$

In (100) and (101) we do the sum member to member

$$2u = (u_1 + k_1) + (u_2 + k_2)$$

that gives

$$u = \frac{1}{2}(u_1 + u_2) + \frac{1}{2}k_1 + \frac{1}{2}k_2$$

Setting $u_0 = \frac{1}{2}(u_1 + u_2)$, $v_1 = \frac{1}{2}k_1$, $v_2 = \frac{1}{2}k_2$ we have:

$$u = u_0 + v_1 + v_2 \tag{102}$$

we replace u by $u_0 + v_1 + v_2$ in $(97)_1$, and we set

$$h = h_0\chi_O + u_0\chi_\omega \in L^2(Q). \tag{103}$$

On can state the following proposition.

Proposition 11. *Let u be in $L^2(\omega \times (0,T))$ and $u_0 \in Y_\lambda + \mathcal{M}$*

$$S(\lambda,\tau)(u) = \int_0^T \int_O hy(\lambda,\tau)dxdt + \int_0^T \int_\omega uy(\lambda,\tau)dxdt$$

then the two proposition are equivalent

1. *find u such that S satisfies (83)-(85)*
2. *find $(v_1, v_2) \in Y_\lambda^\perp \times \mathcal{M}^\perp$ such that if $u = u_0 + v_1 + v_2$ and if q is the solution of (97) then q satisfies (98).*

5. Detection of Parameters

We are now able to give the expression of the sentinel S defined by (82)-(85) and identify the parameter λ_i.

5.1. Expression of the Discriminating Sentinel with Given Sensitivity

We consider the results obtained in the section 3 and we assume that h given by (103) and θ given by Theorem 1 are such that $\theta h \in L^2(O \times (0,T))$. Let $(\widehat{v}_\theta, \widehat{w}_\theta, \widehat{q}_\theta)$ as in Theorem 5. Since $(\widehat{v}_\theta, \widehat{w}_\theta) = (-(\widehat{\rho}_\theta\chi_\omega - P_1\widehat{\rho}_\theta), -(\widehat{\rho}_\theta\chi_\omega -$

$P_2\widehat{\rho}_\theta))$ realizes the minimum in $L^2(\omega \times (0, T))$ among all controls (v, w) such that the pair $((v, w); q)$ satisfies (3)-(6), using (102) , we deduce that $u = u_0 + \widehat{v}_\theta + \widehat{w}_\theta = u_0 - (\widehat{\rho}_\theta\chi_\omega - P_1\widehat{\rho}_\theta) - (\widehat{\rho}_\theta\chi_\omega - P_2\widehat{\rho}_\theta)$. Consequently, replacing u by its expression in (82), the function S becomes:

$$\begin{aligned} S(\lambda, \tau)(u) &= \int_0^T \int_O h_0 y(\lambda, \tau) dx dt \\ &+ \int_0^T \int_\omega (u_0 - (\widehat{\rho}_\theta\chi_\omega - P_1\widehat{\rho}_\theta) - (\widehat{\rho}_\theta\chi_\omega - P_2\widehat{\rho}_\theta)) y(\lambda, \tau) dx dt \end{aligned} \tag{104}$$

and $(u, S(u))$ is such that (82)-(85) holds.

5.2. Identification of Parameter λ_i

Let us show now, how the sentinel permit to detect the pollution λ_i. Let y_0 be the solution of (76) for $\lambda = 0, \tau = 0$. One obtain from (79) that $y_0 = 0$ in Q. Then taking $\lambda = 0, \tau = 0$ in (104), we can calculate

$$S(0,0)(u) = \int_0^T \int_O h_0 y_0 dx dt + \int_0^T \int_\omega (u_0 + \widehat{v}_\theta + \widehat{u}_\theta) y_0 dx dt = 0. \tag{105}$$

From (83) we have

$$S(\lambda, \tau)(u) \simeq S(0,0) + \sum_{i=1}^{M_2} \lambda_i \frac{S}{\partial \lambda_i}(0,0), \text{ for } \lambda_i \text{ and } \tau \text{ small} \tag{106}$$

If y_{obs} is known, its comes from (81) and (84) that

$$S(\lambda, \tau)(u) = \int_0^T \int_O h_0 m_0 dx dt + \int_0^T \int_\omega (u_0 + \widehat{v}_\theta + \widehat{w}_\theta) m_0 dx dt. \tag{107}$$

Thus,

$$S(\lambda, \tau) - S(0,0) = \int_0^T \int_O h_0(m_0 - y_0) dx dt + \int_0^T \int_\omega (u_0 + \widehat{v}_\theta + \widehat{w}_\theta)(m_0 - y_0) dx dt.$$

We have also the following information

$$\sum_{i=1}^{M_1} \lambda_i \frac{S}{\partial \lambda_i}(0,0) = \int_0^T \int_O h_0(m_0 - y_0) dx dt + \int_0^T \int_\omega (u_0 + \widehat{v}_\theta + \widehat{w}_\theta)(m_0 - y_0) dx dt$$

which gives (84) using

$$\sum_{i=1}^{M_1} \lambda_i c_i \approx \int_0^T \int_O h_0(m_0 - y_0)dxdt + \int_0^T \int_\omega (u_0 + \widehat{v}_\theta + \widehat{w}_\theta)(m_0 - y_0)dxdt$$

Now, fixing $i_0 \in \{1, M_1\}$ and choosing $c_{i_0} \neq 0$ and $c_j = 0$, for all $j \in \{1, M_1\}$ with $j \neq i_0$, we obtain of this estimation the parameter λ_{i_0}

$$\begin{aligned} \lambda_{i_0} &\approx \frac{1}{c_{i_0}}\{\int_0^T \int_O h_0(m_0 - y_0)dxdt + \int_0^T \int_\omega (u_0 + \widehat{v}_\theta + \widehat{w}_\theta)(m_0 - y_0)dxdt\} \\ &\approx \frac{1}{c_{i_0}}\{\int_0^T \int_O h_0 m_0 dxdt + \int_0^T \int_\omega (u_0 + \widehat{v}_\theta + \widehat{w}_\theta)m_0 dxdt\} \end{aligned}$$

because $y_0 = 0$ in Q.

REFERENCES

[1] Doubova A., Osses A., Puel J. P. (2002), Exact controllability to trajectories for semilinear heat equations with discontinuous diffusion coefficients. *ESAIM: Control, Optim. Calc. Var.*, 8, 621-661.

[2] Fursikov A., Yu O., Imanuvilov, Controlability of Evolution Equations. *Lecture Notes#34*, Seoul National University, Korea, 1996.

[3] Ainseba B. E., Kernevez J. P. and Luce R., Application des sentinelles à l'identification des pollution dans une rivière. *M2AN Mathematical Modelling an Numerical Analisis.* Vol.28, 3, 1994, pp. 297-312.

[4] Ainseba B. E., Kernevez J. P. and Luce R., Identification de paramètres dans les problèmes non linéaire à donnéés incomplètes. *M2AN Mathematical Modelling an Numerical Analisis.* Vol.28, 3, 1994, pp. 313-328.

[5] Bodart O., sentinels for the identification of an unknown boundary. *Mathematical Models and Methods in Applied Sciences.* Vol.7, N6 (1997) 871-885.

[6] Bodart O., Gonzalez-Burgos M., Perez-Garcia R., Insensitizing control for a heat equation with a nonlinear term involving the state and the gradient. *Nonlinear Analysis*, 57 (2004), 697-711.

[7] Russell D. L., *A unified boundary controllability theory for hyperbolic and parabolic partial differential equations*, N^0.52 (1973), p.189-211.

[8] Fernandez-Cara E., Gonzalez-Burgos M., Guerrero S. et Puel J-P.: Null controllability of the heat equation with boundary Fourier condition: the linear case. *ESAIM: COCV*, Vol.12, N°3, pp.442-465.

[9] Fernández-Cara E. (1997), Nul controllability of the semilinear heat equation, *ESAIM: Control, Optim. Calc. Var.*, 2, 87-103.

[10] Fernández-Cara E., GuerreroS., Global Carleman inequalities for parabolic systems and applications to controllability. *SIAM Journal on control and optimisation*, Vol.45, N°4 (2006), pp.1395-1446.

[11] Fernández-Cara E., Zuazua E. (2000), The cost of approximate controllability for heat equations: the linear case, *Adv. Differ. Equations*, 5, 465-514.

[12] Zuazua E.: Finite dimentional null controlability for the semi-linear heat equation. *J. Math. Pures Appli.* (9)76(1997), 237-264. Zb1 0872.93014 MR 1441986.

[13] Lebeau G., Robbiano L., Contrôle exacte de l'équation de la chaleur, *Comm. partial Differential Equations,* N^0 .20 (1995), p. 335-356.

[14] Massengo Mophou G., Nakoulima O, Sentinel with given sensitivity. *European Journal of Applied Mathematics.* Vol.19, issue 01 (2008) pp. 21-40.

[15] Lions J. L., Sentinelles pour les systèmes distribués à données incomplètes [Sentinels for Distributed Systems data is incomplete]. *Recherches en Mathématiques Appliquées 21,* Masson, Paris 1992.

[16] Lions J. L. and Magenes M., Problemes aux limites non homogènes et applications [Problems with inhomogeneous limits and applications]. *Dunod, Vol.1 et 2*, Paris 1968. Zb1 0165.10801 MR 1159093.

[17] Lions J. L., Controle optimal de systèmes gouvernés par des équations aux dérivées partielles [Optimal control of government systems by partial differential equations]. *Dunod, Gauthier-Villars*, Paris 1968. Zb1 017941801 MR 1159093.

[18] Lions J. L., Contrôle des systèmes distribués singuliers [Controlled Distributed Systems singular]. *BORDAS*, Gauthier-Villars, Paris, 1983, ISBN 2-04-015539-2.

[19] Quelques méthodes de résolution des problèmes aux limites non linéaires [Some methods of solving problems in the non-linear boundaries]. *Dunod*, Gauthier-Villars, Paris 1969

[20] Nakoulima O., A revision of J.-L. Lions' notion of sentinels. *Portugal. Math. Vol. 65, Fasc. 1*, 2008, 1-22.

[21] Yu. Imanuvilov O.,(1995), Controllability of parabolic equations. *Sbornik Mathematics 186*: 6, 879-900.

[22] Cazenave Th. and Haraux A., Introduction aux problèmes d'évolution semi-linéaires [Introduction to the problems of semi-linear evolution]. *Collection Mathématiques et Applications de la Smai Editions Ellipse*, Paris 1991.

[23] Barbu V. (2000), Exact controllability of the superlinear heat equation, *Applied Mathematics and Optimization, 42, no 1*, 73-89.

In: Evolutionary Processes and Applications
Editor: G. M. N'Guérékata
ISBN: 978-1-53616-350-6

Chapter 7

Local Stable Manifolds of Admissible Classes for Parabolic Functional Equations and Applications to Hutchinson Models

***Nguyen Thieu Huy*[1,*] *and Dinh Xuan Khanh*[2,†]**
[1]School of Applied Mathematics and Informatics,
Hanoi University of Science and Technology,
Dai Co Viet, Hanoi, Vietnam
[2]Faculty of Mathematics, Hai Phong University Phan Dang Luu,
Kien An, Hai Phong, Vietnam

Abstract

We prove the existence of local stable manifolds of admissible classes for solutions to non-autonomous parabolic functional equation of the form $\dot{u}(t) = A(t)u(t) + f(t, u_t)$, $t \geq 0$. The special feature of this kind of manifolds is that they are constituted by trajectories of the solutions belonging to admissible function spaces. The existence of these manifolds obtained in the case that the family of operators $(A(t))_{t\geq 0}$ generate the evolution family $(U(t,s))_{t\geq s\geq 0}$ having an exponential dichotomy on the half-line and the nonlinear forcing term f satisfies the local φ-Lipschitz condition, i.e., $\|f(t,\phi) - f(t,\psi)\| \leq \varphi(t)\|\phi - \psi\|_{\mathcal{C}}$ where ϕ, ψ belong

[*]Corresponding Author's Email: huy.nguyenthieu@hust.edu.vn.
[†]Corresponding Author's Email: khanhdx.dhhp@gmail.com.

to some ball in $\mathcal{C} := C([-r,0],X)$, $\varphi(\cdot)$ is in some admissible Banach function space and satisfies some certain conditions. We then apply our results to Hutchinson model to show the local admissibly stable manifold for Hutchinson equations.

Keywords: exponential dichotomy, partial functional differential equations, admissibility of function spaces, local stable manifolds of admissible classes

1. INTRODUCTION

Let consider the semi-linear functional differential equations of the form

$$\frac{du}{dt} = A(t)u(t) + f(t,u_t), \quad t \in [0,+\infty), \tag{1.1}$$

where $A(t)$ is a (possibly unbounded) linear operator on a Banach space X for every fixed t; $f : \mathbb{R}_+ \times \mathcal{C} \to X$ is a continuous nonlinear operator with $\mathcal{C} := C([-r,0],X)$; and u_t is the history function defined by $u_t(\theta) := u(t+\theta)$ for $\theta \in [-r,0]$.

In the present literature, for the existence of integral (stable, unstable, or center) manifolds ones need the sufficient condition for the that the family of linear operators $(A(t))_{t\geq 0}$ (the linear part) generates an evolution family having an exponential dichotomy or trichotomy; and for the nonlinear term f, the most popular condition is its Lipschitz continuity with a sufficiently small Lipschitz constant, i.e., $\|f(t,\phi) - f(t,\psi)\| \leq q\|\phi - \psi\|_{\mathcal{C}}$ for q small enough (see [1], [2], [3], [4], [5], [7], [8], [9], [10]). However, for equations arising in complicated reaction-diffusion processes, the mapping f represents the source of material or population whose Lipschitz coefficients, in many contexts, depend on time (see [11, Chapt. 11], [12], [13]) in diversified manners. Therefore, one may not hope that the mapping f is Lipschitz continuous with small Lipschitz constant. Moreover, the manifolds in the literature are created by trajectories of bounded solutions (or rescaledly bounded solutions) on the positive half-line or the whole line. We refer reader to [1], [2] [3], [9] and references therein for more information on this matter.

Recently, for the partial differential equations *without delay* we have proved (in [14]) the existence a new type of invariant manifolds, namely, the *invariant manifold of admissible classes*. These manifolds are constituted of trajectories of the solutions belonging to admissible function spaces which contain wide classes of function spaces like L_p-spaces and many other function spaces occurring in interpolation theory such as the Lorentz spaces $L_{p,q}$. We have used the Lyapunov-Perron equation to construct the mild solutions belonging an admissible space on which one could employ some well-known principles in mathematical analysis such as the contraction mapping principle, the implicit function theorem, cone inequality etc. The use of admissible spaces has helped us to construct such invariant manifolds when the mapping f has the Lipschitz coefficient depending on time and not small in classical sense. In fact, we have imposed a more general condition on the nonlinear term f, namely, the φ-Lipschitz condition, i.e.,, $\|f(t,\phi) - f(t,\psi)\| \leq \varphi(t)\|\phi - \psi\|$ for $\varphi(t)$ being a real and positive function which belongs to the admissible Banach function space. Consequently, we obtain the existence of such invariant manifolds of admissible class for evolution equations *without delay* in [14]. And that result has been extended to more general case for equations defined on whole-line (see [15]).

The purpose of the present chapter is to extend the results in [14] to the case of *partial differential functional equations with finite delay* (PFDE). The difficulties when passing to the case of PFDE are lying in the facts that each linear operator $A(t)$, $t \geq 0$, does not act on the same Banach space as that the surfaces of the manifold belong to (in fact, $A(t)$ acts on X and the latter belong to $\mathcal{C}$). Moreover, since the nonlinear term f is φ-Lipschitz, the existence and uniqueness theorem has not been available. To overcome such difficulties, we reformulate the definition of invariant manifolds of admissible class such that it contains the existence and uniqueness theorem as a property of the manifold (see Definition 2.3 below). Furthermore, we combine the dichotomy estimates, Hölder type inequalities, and admissibility of function spaces in such a way that it allows to construct the form of bounded solutions represented by Lyapunov-Perron equation used in setting of local stable manifolds of admissible class. Our main results are contained in Lemma 2.4 and Theorems 2.6 and 2.7. We then apply our obtained results to prove the existence of an invariant local stable manifold of admissible class for the Hutchinson Models.

We now recall some notions.

For a Banach space X (with a norm $\|\cdot\|$) and a given $r > 0$ we denote by $\mathcal{C} := C([-r,0],X)$ the Banach space of all continuous functions from $[-r,0]$ into X, equipped with the norm $\|\phi\|_{\mathcal{C}} = \sup_{\theta\in[-r,0]}\|\phi(\theta)\|$ for $\phi \in \mathcal{C}$. For a continuous function $u : [-r,\infty) \to X$, the *history function* $u_t \in \mathcal{C}$ is defined by $u_t(\theta) = u(t+\theta)$ for all $\theta \in [-r,0]$.

Definition 1.1. A family of bounded linear operators $(U(t,s))_{t\geq s\geq 0}$ on a Banach space X is a *(strongly continuous, exponential bounded) evolution family* if

(i) $U(t,t) = Id$ and $U(t,r)U(r,s) = U(t,s)$ for all $t \geq r \geq s \geq 0$,

(ii) the map $(t,s) \mapsto U(t,s)x$ is continuous for every $x \in X$,

(iii) there are constants $K, c \geq 0$ such that $\|U(t,s)x\| \leq Ke^{c(t-s)}\|x\|$ for all $t \geq s \geq 0$ and $x \in X$.

We refer the readers to Pazy [23] for more details on the notion of evolution families and their applications to partial differential equations (see also Nagel and Nickel [22] for a detailed discussion of well-posedness for non-autonomous abstract Cauchy problems on the whole line $\mathbb{R}$). Roughly speaking, an evolution family represents the form of solutions to the well-posed, non-autonomous abstract Cauchy problem

$$\begin{cases} \frac{du}{dt} &= A(t)u(t), \quad t \geq s \geq 0, \\ u(s) &= x \end{cases} \tag{1.2}$$

i.e., for the well-posed Cauchy problem (1.2) there exists an evolution family $(U(t,s))_{t\geq s\geq 0}$ such that the solution to (1.2) is given by $u(t) = U(t,s)u(s)$ for every $u(s)$ belonging domain of the linear operator $A(s)$. In that case, we formally say that this evolution family $(U(t,s))_{t\geq s\geq 0}$ is generated by the family $(A(t))_{t\geq 0}$.

We next briefly recall some notions on function spaces taken from Massera and Schäffer [20] and Huy et al. [14, 15, 19].

Let E be admissible function spaces and E' be its associate space defined as in [15, 19]. Then, we set

$$\mathcal{E} := \mathcal{E}(\mathbb{R}, \mathcal{C}) := \{f : \mathbb{R} \to \mathcal{C} : f \text{ is strongly measurable and } \|f(\cdot)\|_{\mathcal{C}} \in E\} \tag{1.3}$$

endowed with the norm

$$\|f\|_{\mathcal{E}} := \|\|f(\cdot)\|_{\mathcal{C}}\|_E.$$

Then, clearly $\mathcal{E}$ is a Banach space, called *the Banach space corresponding to the admissible function space E*.

We refer the readers to [15] for various examples of admissible spaces and their applications to invariant manifolds of admissible classes. Typical examples of admissible spaces satisfying the above hypothesis are L_p-spaces with one type of exponentially L_p-invariant functions of the form $\beta e^{-\alpha|t|}$ for $t \in \mathbb{R}$ and any fixed $\beta, \alpha > 0$.

Remark 1.2. Let E be an admissible Banach function space and E' be its associate space. Then, from [20, Chapt. 2, 22.M] we also have that the following "Hölder's inequality" holds:

$$\int_0^\infty |\varphi(t)\psi(t)|dt \leq \|\varphi\|_E \|\psi\|_{E'} \quad \text{for all } \varphi \in E,\ \psi \in E'. \tag{1.4}$$

In order to study the local-stable manifolds of $\mathcal{E}$-class for partial functional differential equations we need some restrictions on the admissible Banach function spaces and assume the following hypothesis.

Hypothesis 1.3. In this chapter we will consider the admissible Banach function space E such that its associate space E' is also an admissible Banach function space. Moreover, for such an admissible Banach function space E we suppose that E' contains an *exponentially E-invariant* function, that is the function $\varphi \geq 0$ and $\varphi \in E'$ having the property that, for any fixed $\nu > 0$ the function h_ν defined by

$$h_\nu(t) := \|e^{-\nu|t-\cdot|}\varphi(\cdot)\|_{E'} \quad \text{for } t \geq 0$$

belongs to E.

We also give here some examples of the admissible Banach function spaces and their associate function spaces which satisfy the above Hypothesis. We note that the functions $\varphi(t) = \beta e^{-\alpha t}$ for $t \geq 0$ and fixed $\beta, \alpha > 0$ are exponentially E-invariant to any admissible Banach function space E.

Example 1.4. $L'_p(\mathbb{R}_+) = L_q(\mathbb{R}_+)$ for $\frac{1}{p} + \frac{1}{q} = 1,\ 1 \leq p \leq \infty$ and the Lorentz spaces $L_{p,q}(\mathbb{R}_+)$ have $L'_{p,q}(\mathbb{R}_+) = L_{p',q'}(\mathbb{R}_+)$ with $\frac{1}{p}+\frac{1}{p'} = 1,\ \frac{1}{q}+\frac{1}{q'} = 1,\ 1 < p < \infty,\ 1 \leq q \leq \infty$. Besides the functions of the form $\varphi(t) = \beta e^{-\alpha t}$, one can see that the functions of the form $\varphi = c\chi_{[a,b]}$ for any fixed constant $c > 0$ and any finite interval $[a, b] \subset \mathbb{R}_+$ are also exponentially E-invariant functions, in which E are the spaces $L_p(\mathbb{R}_+)$ or $L_{p,q}(\mathbb{R}_+)$.

2. Exponential Dichotomy and Local-Stable Manifold of $\mathcal{E}$-Class

In this section, we will prove the existence of local-stable manifold of $\mathcal{E}$-class (see Definition 2.3 below) for the solutions to Eq. (1.1). Throughout this section we assume that the evolution family $(U(t,s))_{t\geq s\geq 0}$ has an exponential dichotomy on $\mathbb{R}_+$. We now make precisely the notion of exponential dichotomies in the following definition.

Definition 2.1. An evolution family $(U(t,s))_{t\geq s\geq 0}$ on the Banach space X is said to have an *exponential dichotomy* on $[0,\infty)$ if there exists bounded linear projections $P(t)$, $t\geq 0$, on X and positive constants N, ν such that

(a) $U(t,s)P(s) = P(t)U(t,s), \quad t\geq s\geq 0,$

(b) the restriction $U(t,s)_| : \mathrm{Ker}P(s) \to \mathrm{Ker}P(t)$, $t\geq s\geq 0$, is an isomorphism, and we denote its inverse by $U(s,t)_| := (U(t,s)_|)^{-1}$, $0\leq s\leq t$,

(c) $\|U(t,s)x\| \leq Ne^{-\nu(t-s)}\|x\|$ for $x\in P(s)X$, $t\geq s\geq 0$,

(d) $\|U(s,t)_| x\| \leq Ne^{-\nu(t-s)}\|x\|$ for $x\in \mathrm{Ker}P(t)$, $t\geq s\geq 0$.

The projections $P(t)$, $t\geq 0$, are called the *dichotomy projections*, and the constants N, ν - the *dichotomy constants*.

Using the exponential dichotomy of $(U(t,s))_{t\geq s\geq 0}$ we can define the family of operators $(\widetilde{P}(t))_{t\geq 0}$ on $\mathcal{C}$ as follows.

$$\begin{aligned} \widetilde{P}(t) : \mathcal{C} &\to \mathcal{C} \\ (\widetilde{P}(t)\phi)(\theta) &= U(t-\theta,t)P(t)\phi(0) \text{ for all } \theta\in[-r,0]. \end{aligned} \tag{2.1}$$

Then, we have that $(\widetilde{P}(t))^2 = \widetilde{P}(t)$, and therefore the operators $\widetilde{P}(t)$, $t\geq 0$ are projections on $\mathcal{C}$. Moreover, $\mathrm{Im}\widetilde{P}(t) = \{\phi\in\mathcal{C} : \phi(\theta) = U(t-\theta,t)\nu_0 \ \forall\theta\in[-r,0] \text{ for some } \nu_0\in \mathrm{Im}P(t)\}$.

We then recall the notion of the φ-Lipschitz property of the nonlinear term f.

Definition 2.2. Let V be an admissible Banach function space and φ be a positive function belonging to V, and B_ρ be the ball with radius ρ centered at the origin in X, i.e., $B_\rho := \{x\in X : \|x\|\leq\rho\}$. A function $f : [0,\infty)\times B_\rho \to X$ is said to belong to class (M,ϕ,ρ) for some positive conditions M,ϕ,ρ if it satisfies

(i) $f(t, \phi) \le M\varphi(t)$ for all $t \in \mathbb{R}_+$ and $\phi \in \mathcal{B}_\rho$

(ii) $\|f(t, \phi_1) - f(t, \phi_2)\| \le \varphi(t)\|\phi_1 - \phi_2\|_{\mathcal{C}}$ for all $t \in \mathbb{R}_+$ and all $\phi_1, \phi_2 \in \mathcal{B}_\rho$.

Note that if $f(t, 0) = 0$ then the condition (ii) in the above definition already implies that f belongs to class (ρ, ϕ, ρ).

Next, by by mild solutions of Eq. (1.1) we mean the solution of the following integral equation

$$\begin{cases} u(t) &= U(t,s)u(s) + \int_s^t U(t,\xi)f(\xi, u_\xi)d\xi \quad \text{for } t \ge s \ge 0, \\ u_s &= \phi \in \mathcal{C}, \end{cases} \tag{2.2}$$

with $(U(t, s))_{t\ge s\ge 0}$ is an evolution family generated by the well-posed Cauchy problem (1.2). We refer to J. Wu [26] for more detailed discussions on the mild solutions of partial functional differential equations and their relations with classical solutions.

Let now $\mathcal{E} := \mathcal{E}(\mathbb{R}_+, \mathcal{C})$ be the admissible Banach space corresponding to admissible Banach function space E. We then give the definition of a local - stable manifold of $\mathcal{E}$-class for the solutions of Eq. (2.2).

Definition 2.3. A set $S \subset \mathbb{R}_+ \times \mathcal{C}$ is said to be an *invariant local - stable manifold of $\mathcal{E}$-class* for the solutions to Eq. (2.2) if for every $t \in \mathbb{R}_+$ the phase space $\mathcal{C}$ splits into a direct sum $\mathcal{C} = \widetilde{X}_0(t) \oplus \widetilde{X}_1(t)$ with corresponding projections $\widetilde{P}(t)$ (i.e., $\widetilde{X}_0(t) = \mathrm{Im}\widetilde{P}(t)$ and $\widetilde{X}_1(t) = \mathrm{Ker}\widetilde{P}(t)$) such that

$$\sup_{t\ge 0} \|\widetilde{P}(t)\| < \infty,$$

and there exists positive constants ρ, ρ_0, ρ_1, and a family of Lipschitz mappings

$$\Phi_t : B_{\rho_0} \cap \widetilde{X}_0(t) \to B_{\rho_1} \cap \widetilde{X}_1(t), \quad t \in \mathbb{R}_+$$

with the Lipschitz constants independent of t such that

(i) $S = \{(t, \psi + \Phi_t(\psi)) \in \mathbb{R}_+ \times (\widetilde{X}_0(t) \oplus \widetilde{X}_1(t)) \mid t \in \mathbb{R}_+, \psi \in B_{\rho_0} \cap \widetilde{X}_0(t)\}$, and we denote

$$S_t := \{\psi + \Phi_t(\psi) : (t, \psi + \Phi_t(\psi)) \in S\},$$

(ii) S_t is homeomorphic to $B_{\rho_0} \cap \widetilde{X}_0(t)$ for all $t \ge 0$,

(iii) to each $\phi \in S_s$ there corresponds one and only one solution $u(t)$ to Eq. (2.2) on $[s-r,\infty)$ satisfying the initial condition $u_s = \phi$ and the function $\chi_{[s,\infty)}(t)u_t,\ t \geq 0$, belongs to the ball with radius ρ in $\mathcal{E}$ where $\chi_{[s,\infty)}$ is the characteristic function of the interval $[s,\infty)$. Moreover, any two solutions $u(t)$ and $v(t)$ of Eq. (2.2) corresponding to different functions $\phi_1, \phi_2 \in S_s$ attract each other exponentially in the sense that, there exists positive constants μ and C_μ independent of $s \geq 0$ such that

$$\|u_t - v_t\|_C \leq C_\mu e^{-\mu(t-s)}\|(\widetilde{P}(s)\phi_1)(0) - (\widetilde{P}(s)\phi_2)(0)\| \text{ for } t \geq s, \quad (2.3)$$

(iv) S is positively invariant under Eq. (2.2), i.e., if $u(t), t \geq s-r$, is a solution to Eq. (2.2) such that initial condition $u_s \in S_s$ and $\chi_{[s,\infty)}(t)u_t \in \mathcal{B}_\rho \subset \mathcal{E}$ then we have $u_t \in S_t$ for all $t \geq s$.

Note that if we identify $\widetilde{X}_0(t) \oplus \widetilde{X}_1(t)$ with $\widetilde{X}_0(t) \times \widetilde{X}_1(t)$, then we can write $S_t = graph(\Phi_t)$. This definition is local version of the definition of invariant stable manifolds in the papers [18, 9, 14].

Let the evolution family $(U(t,s))_{t\geq s\geq 0}$ have an exponential dichotomy with the dichotomy projections $P(t),\ t \geq 0$, and constants $N,\ \nu > 0$. Note that the exponential dichotomy of $(U(t,s))_{t\geq s\geq 0}$ implies that $H := \sup_{t\geq 0}\|P(t)\| < \infty$ and the map $t \mapsto P(t)$ is strongly continuous (see [21, Lemma 4.2]). We can then define the Green's function on the half-line as follows

$$\mathcal{G}(t,\tau) = \begin{cases} P(t)U(t,\tau) & \text{for } t > \tau \geq 0 \\ -U(t,\tau)_|(I - P(\tau)) & \text{for } 0 \leq t < \tau. \end{cases} \quad (2.4)$$

It follows from the exponential dichotomy of $(U(t,s))_{t\geq s\geq 0}$ that

$$\|\mathcal{G}(t,\tau)\| \leq N(1+H)e^{-\nu|t-\tau|} \quad \text{for all } t \neq \tau.$$

The following lemma gives the form of solution to Eq. (2.2) which belongs to $\mathcal{E}$.

Lemma 2.4. *Let the evolution family $(U(t,s))_{t\geq s\geq 0}$ have an exponential dichotomy with the dichotomy projections $P(t),\ t \geq 0$ and constants $N,\ \nu > 0$. Let E be an admissible Banach function space, E' be its associate space and $\mathcal{E} := \mathcal{E}(\mathbb{R}_+, \mathcal{C})$ be admissible Banach space corresponding to E. Suppose that $\varphi \in E'$ is an exponentially E-invariant function defined as in Hypothesis 1.3. Let $f : \mathbb{R}_+ \times B_\rho \to X$ belong to class (M, ϕ, ρ) for some positive* *constants M, ρ. And let $u(t)$ be a solution to Eq. (2.2) such that the function $\chi_{[s,\infty)}(t)u_t,\ t \geq 0$, belongs to $\mathcal{B}_\rho \subset \mathcal{E}$ for fixed $s \geq 0$. Then, for $t \geq s$ we can rewrite $u(t)$ in the form*

$$\begin{cases} u(t) = U(t,s)\nu_0 + \displaystyle\int_s^\infty \mathcal{G}(t,\tau)f(\tau,u_\tau)d\tau, \\ u_s = \phi \in \mathcal{C} \end{cases} \tag{2.5}$$

for some $\nu_0 \in X_0(s) = P(s)X$, where $\mathcal{G}(t,\tau)$ is the Green's function defined as in (2.4).

Proof. Set $z(t) := \chi_{[s,\infty)}(t)u_t$, $t \geq 0$, and put $y(t) = \displaystyle\int_s^\infty \mathcal{G}(t,\tau)f(\tau,u_\tau)d\tau$ for $t \geq s$ and $y(t) = 0$ for $0 \leq t < s$. We have

$$\|y(t)\| \leq N(1+H)\int_s^\infty e^{-\nu|t-\tau|}\varphi(\tau)\|u_\tau\|_{\mathcal{C}}d\tau = N(1+H)\int_0^\infty e^{-\nu|t-\tau|}\varphi(\tau)\|z(\tau)\|_{\mathcal{C}}d\tau,\ t \geq 0.$$

Using "Hölder's inequality" (1.4) we obtain

$$\|y(t)\| \leq N(1+H)\|e^{-\nu|t-\cdot|}\varphi(\cdot)\|_{E'}\|z\|_{\mathcal{E}},\ t \geq 0. \tag{2.6}$$

Note that the function $h_\nu(t) = \|e^{-\nu|t-\cdot|}\varphi(\cdot)\|_{E'}$, $t \geq 0$, belongs to E. Therefore, by Banach lattice property we have that the function $\|y(\cdot)\| \in E$. Similarly, we also have the function $\|u(\cdot)\| \in E$. On the other hand,

$$\begin{aligned} U(t,s)y(s) &= -\int_s^t U(t,s)U(s,\tau)_|(I-P(\tau))f(\tau,u_\tau)d\tau \\ &\quad -\int_t^\infty U(t,s)U(s,\tau)_|(I-P(\tau))f(\tau,u_\tau)d\tau \\ &= -\int_s^t U(t,\tau)(I-P(\tau))f(\tau,u_\tau)d\tau - \int_t^\infty U(t,\tau)_|(I-P(\tau))f(\tau,u_\tau)d\tau. \end{aligned}$$

Therefore,

$$y(t) = U(t,s)y(s) + \int_s^t U(t,\tau)f(\tau,u_\tau)d\tau.$$

Since $u(t)$ is a solution of Eq. (2.2) we obtain that $u(t)-y(t) = U(t,s)(u(s)-y(s))$. Put now $\nu_0 = u(s)-y(s)$. From the functions $\|y(t)\|, \|u(t)\|$ belong to E implies that $\nu_0 \in X_0(s)$ and $P(s)u(s) = P(s)\phi(0) = \nu_0$. Therefore, $u(t) = U(t,s)\nu_0 + y(t)$ for $t \geq s$. □

Remark 2.5. The Eq. (2.5) is called the *Lyapunov-Perron equation*. By computing directly, we can see that the converse of Lemma 2.4 is also true. This means that, all solutions of the integral equation (2.5) belonging to $\mathcal{E}$ also are solutions of Eq. (2.2) in admissible Banach space $\mathcal{E}$ for $t \geq s$.

Using admissibility, we construct the structure of solutions of Eq. (2.2) in the next theorem. To do this, we first denote by $\mathbb{B}_l := \{\phi \in \mathcal{C} : \|\phi\|_{\mathcal{C}} \leq l\}$ the ball with radius l in $\mathcal{C}$.

Theorem 2.6. *Under the hypotheses of Lemma 2.4 we let $\varphi \in E'$ be an exponentially E-invariant function and the corresponding function h_ν defined as in Hypothesis 1.3. Then, we have that if the function f belongs to the class (M, ϕ, ρ) with positive numbers M, ρ such that*

$$N(1+H)e^{\nu r}\|h_\nu\|_E < \frac{1}{2},$$

then for $l = \frac{\rho}{2NN_1e^{\nu r}\|e_\nu\|}$ there corresponds to each $\phi \in \mathbb{B}_l \cap \mathrm{Im}\widetilde{P}(s)$ one and only one solution $u(t)$ of Eq. (2.2) on $[s-r, \infty)$ satisfying the condition $\widetilde{P}(s)u_s = \phi$ and the function $\chi_{[s,\infty)}(t)u_t,\ t \geq 0$, belongs to the ball $\mathcal{B}_\rho$ in $\mathcal{E}$. Moreover, if $\frac{N(1+H)e^{\nu r}}{1-e^{-(\nu-\mu)}}(N_1\|\Lambda_1T_1^+\varphi\|_\infty + N_2\|\Lambda_1\varphi\|_\infty) < 1$, then the following estimate is valid for any two solutions $u(t), v(t)$ corresponding to different initial functions $\phi_1, \phi_2 \in \mathbb{B}_l \cap \mathrm{Im}\widetilde{P}(s)$:

$$\|u_t - v_t\|_{\mathcal{C}} \leq C_\mu e^{-\mu(t-s)}\|\phi_1(0) - \phi_2(0)\| \qquad \text{for all } t \geq s \geq 0, \qquad (2.7)$$

where μ is a positive number satisfying

$$0 < \mu < \nu + \ln\left(1 - N(1+H)e^{\nu r}(N_1\|\Lambda_1T_1^+\varphi\|_\infty + N_2\|\Lambda_1\varphi\|_\infty)\right),$$

and

$$C_\mu := \frac{Ne^{\nu r}}{1 - \frac{N(1+H)e^{\nu r}}{1-e^{-(\nu-\mu)}}(N_1\|\Lambda_1T_1^+\varphi\|_\infty + N_2\|\Lambda_1\varphi\|_\infty)}.$$

Proof. To show the existence and uniqueness of solution for Eq. (2.2) with each $\phi \in \mathbb{B}_l \cap \mathrm{Im}\widetilde{P}(s)$. Denote $C_b([s-r,\infty), X)$ is the set of bounded, continuous and X-valued functions defined on $[s-r,\infty)$. Setting $\nu_0 := \phi(0)$, for each $z \in \mathcal{B}_\rho \subset \mathcal{E} := \mathcal{E}(\mathbb{R}_+, \mathcal{C})$ then we define the function $H : \mathcal{B}_\rho \to C_b([s-r,\infty), X)$ as follows

$$(Hz)(t) = \begin{cases} U(2s-t, s)\nu_0 + \displaystyle\int_s^\infty \mathcal{G}(2s-t,\tau)f(\tau, z(\tau))d\tau & \text{for } s-r \leq t \leq s, \\ U(t,s)\nu_0 + \displaystyle\int_s^\infty \mathcal{G}(t,\tau)f(\tau, z(\tau))d\tau & \text{for } t \geq s. \end{cases}$$

We will prove that the transformation T defined by

$$(Tz)(t) = \begin{cases} (Hz)_t & \text{for } t \geq s, \\ 0 & \text{for } 0 \leq t < s \end{cases}$$

acts from $\mathcal{B}_\rho$ into $\mathcal{B}_\rho$ and is a contraction mapping. In fact, we have

$$\|(Hz)(t)\| \le \begin{cases} Ne^{-\nu(s-t)}\|\nu_0\| + N(1+H)\int_s^\infty e^{-\nu|2s-t-\tau|}\varphi(\tau)\|z(\tau)\|_{\mathcal{C}}d\tau & \text{for } s-r \le t \le s, \\ Ne^{-\nu(t-s)}\|\nu_0\| + N(1+H)\int_s^\infty e^{-\nu|t-\tau|}\varphi(\tau)\|z(\tau)\|_{\mathcal{C}}d\tau & \text{for } t \ge s. \end{cases}$$

Therefore, for $t \ge s$ then

$$\|(Hz)_t\|_{\mathcal{C}} = \sup_{\theta\in[-r,0]}\|(Hz)(t+\theta)\| \le Ne^{\nu r}e^{-\nu(t-s)}\|\nu_0\| + N(1+H)e^{\nu r}\int_s^\infty e^{-\nu|t-\tau|}\varphi(\tau)\|z(\tau)\|_{\mathcal{C}}d\tau.$$

Using “Hölder’s inequality” (1.4) we obtain

$$\|(Hz)_t\|_{\mathcal{C}} \le Ne^{\nu r}T_s^+e_\nu(t)\|\nu_0\| + N(1+H)e^{\nu r}h_\nu(t)\|z\|_{\mathcal{E}},$$

where $h_\nu(t) = \|e^{-\nu|t-\cdot|}\varphi(\cdot)\|_{E'}$ and $T_s^+e_\nu(t) = \begin{cases} e^{-\nu(t-s)} & \text{for } t \ge s, \\ 0 & \text{for } 0 \le t < s \end{cases}.$

Since the functions $T_s^+e_\nu(\cdot),\ h_\nu(\cdot) \in E$ and by the Banach lattice property of E it follows that $\|(Tz)(\cdot)\|_{\mathcal{C}} \in E$ and

$$\|Tz\|_{\mathcal{E}} \le NN_1e^{\nu r}\|e_\nu\|_E\|\nu_0\| + N(1+H)e^{\nu r}\|h_\nu\|_E\|z\|_{\mathcal{E}} < \rho.$$

Thus, $Tz \in \mathcal{B}_\rho$.

Next, we prove that T is a contraction. For $x, z \in \mathcal{E}$ we have estimate,

$$\|(Hx)_t-(Hz)_t\|_{\mathcal{C}} \le N(1+H)e^{\nu r}\int_s^\infty e^{-\nu|t-\tau|}\varphi(\tau)\|x(\tau)-z(\tau)\|_{\mathcal{C}}d\tau \quad \text{for } t \ge s.$$

By “Hölder’s inequality” (1.4), we obtain

$$\|(Hx)_t-(Hz)_t\|_{\mathcal{C}} \le N(1+H)e^{\nu r}h_\nu(t)\|x-z\|_{\mathcal{E}}.$$

By Banach lattice property of E, we have

$$\|Tx-Tz\|_{\mathcal{E}} \le N(1+H)e^{\nu r}\|h_\nu\|_E\|x-z\|_{\mathcal{E}} < \frac{1}{2}\|x-z\|_{\mathcal{E}}.$$

Thus, T is a contraction. Hence, there exists unique $z \in \mathcal{B}_\rho$ such that $Tz = z$. Therefore, we have

$$z(t) = \begin{cases} (Hz)_t & \text{for } t \ge s, \\ 0 & \text{for } 0 \le t < s. \end{cases}$$

This yields

$$(Hz)(t) = \begin{cases} U(2s-t,s)\nu_0 + \displaystyle\int_s^\infty \mathcal{G}(2s-t,\tau)f(\tau,(Hz)_\tau)d\tau & \text{for } s-r \le t \le s, \\ U(t,s)\nu_0 + \displaystyle\int_s^\infty \mathcal{G}(t,\tau)f(\tau,(Hz)_\tau)d\tau & \text{for } t \ge s. \end{cases}$$

So, $u(t) = (Hz)(t)$ is unique solution of Eq. (2.5). By Lemma 2.4 and Remark 2.5 then $u(t)$ is also unique solution of Eq. (2.2) with $u_s(\theta) = U(s-\theta,s)\nu_0 + \int_s^\infty \mathcal{G}(s-\theta,\tau)f(\tau,u_\tau)d\tau$ for $\theta \in [-r,0]$, and $P(s)u(s) = \nu_0 = \phi(0)$. Therefore, $\widetilde{P}(s)u_s = \phi$ by definition of $\widetilde{P}(s)$ (see Equality (2.1)).

We last show the inequality 2.7. To do this, let $u(t), v(t)$ be the two solutions to Eq. (2.5) corresponding to different initial functions $\phi_1, \phi_2 \in \mathbb{B}_l \cap \mathrm{Im}\widetilde{P}(s)$, respectively. Putting $\nu_1 := \phi_1(0)$, $\nu_2 := \phi_2(0)$ we have that

$$\|u(t)-v(t)\| \le \begin{cases} Ne^{-\nu(t-s)}\|\nu_1-\nu_2\| + N(1+H)\int_s^\infty e^{-\nu|t-\tau|}\varphi(\tau)\|u_\tau - v_\tau\|_C d\tau \text{ if } t \ge s, \\ Ne^{-\nu(s-t)}\|\nu_1-\nu_2\| + N(1+H)\int_s^\infty e^{-\nu|2s-t-\tau|}\varphi(\tau)\|u_\tau - v_\tau\|_C d\tau \text{ if } s-r \le t \le s. \end{cases}$$

Since $t+\theta \in [-r+t,t]$ for fixed $t \in [s,\infty)$ and $\theta \in [-r,0]$, we obtain

$$\|u_t - v_t\|_C \le Ne^{\nu r}e^{-\nu(t-s)}\|\nu_1-\nu_2\| + N(1+H)e^{\nu r}\int_s^\infty e^{-\nu|t-\tau|}\varphi(\tau)\|u_\tau - v_\tau\|_C\, d\tau, \quad t \ge s.$$

Put $h(t) = \begin{cases} \|u_t - v_t\|_C & \text{for } t \ge s, \\ 0 & \text{for } 0 \le t < s \end{cases}$. Then, $h(t) \in E$ and

$$h(t) \le Ne^{\nu r}e^{-\nu(t-s)}\|\nu_1-\nu_2\| + N(1+H)e^{\nu r}\int_s^\infty e^{-\nu|t-\tau|}\varphi(\tau)h(\tau)d\tau, t \ge s \tag{2.8}$$

We will use the cone inequality theorem (see [16, Theorem 2.8]) for admissible Banach function space E and the cone $\mathcal{K}$ being the set of all non-negative functions. We then consider the linear operator A defined for $g \in E$ by

$$(Ag)(t) = \begin{cases} N(1+H)e^{\nu r}\int_s^\infty e^{-\nu|t-\tau|}\varphi(\tau)g(\tau)d\tau & \text{for } \quad t \ge s, \\ 0 & \text{for } \quad 0 \le t < s. \end{cases}$$

By "Hölder's inequality" (1.4) then

$$|(Ag)(t)| \le N(1+H)e^{\nu r}h_\nu(t)\|g\|_E.$$

By the Banach lattice property of E, we have $\|Ag\|_E \le N(1+H)e^{\nu r}\|h_\nu\|_E\|g\|_E$. Therefore, $A \in \mathcal{L}(E)$ and $\|A\| \le N(1+H)e^{\nu r}\|h_\nu\|_E < 1$

Obviously, the cone $\mathcal{K}$ is invariant under the operator A. The inequality (2.8) can now be rewritten by

$$h \leq Ah + z \text{ for } z(t) = Ne^{\nu r}e^{-\nu(t-s)}\|\nu_1 - \nu_2\|.$$

By the cone inequality theorem [16, Theorem 2.8] we obtain that $h \leq g$, where g is a solution in E of the equation $g = Ag + z$ which can be rewritten as

$$g(t) = Ne^{\nu r}e^{-\nu(t-s)}\|\nu_1 - \nu_2\| + N(1+H)e^{\nu r}\int_s^\infty e^{-\nu|t-\tau|}\varphi(\tau)g(\tau)d\tau, t \geq s.$$

We put $w(t) = e^{\mu(t-s)}g(t)$ for $t \geq s$. Then, we obtain that w is solution of the equation

$$w(t) = Ne^{\nu r}e^{-(\nu-\mu)(t-s)}\|\nu_1 - \nu_2\| + N(1+H)e^{\nu r}\int_s^\infty e^{-\nu|t-\tau|+\mu(t-\tau)}\varphi(\tau)w(\tau)d\tau. \tag{2.9}$$

We will find w in $L_\infty[s,\infty)$ which is space of real-valued functions defined and essentially bounded on $[s,\infty)$ (endowed with the sup-norm denoted by $\|\cdot\|_\infty$). We consider the linear operator K defined on $L_\infty[s,\infty)$ as

$$(K\phi)(t) = N(1+H)e^{\nu r}\int_s^\infty e^{-\nu|t-\tau|+\mu(t-\tau)}\varphi(\tau)\phi(\tau)d\tau \text{ for all } t \geq s.$$

By [14, Proposition 2.7], one can easily see that $K \in \mathcal{L}(L_\infty[s,\infty))$ and

$$\|K\| \leq \frac{N(1+H)e^{\nu r}}{1 - e^{-(\nu-\mu)}}(N_1\|\Lambda_1 T_1^+\varphi\|_\infty + N_2\|\Lambda_1\varphi\|_\infty).$$

The equation (2.9) can be rewritten as

$$w = Kw + \tilde{z} \quad \text{for} \quad \tilde{z}(t) = Ne^{\nu r}e^{-(\nu-\mu)(t-s)}\|\nu_1 - \nu_2\|.$$

We have $\|K\| < 1$ if $0 < \mu < \nu + \ln\left(1 - N(1+H)e^{\nu r}(N_1\|\Lambda_1 T_1^+\varphi\|_\infty + N_2\|\Lambda_1\varphi\|_\infty)\right)$. Under this condition, the equation $w = Kw + \tilde{z}$ has the unique solution $w \in L_\infty[s,\infty)$ and $w = (I-K)^{-1}\tilde{z}$. Hence, we obtain that

$$\begin{aligned}\|w\|_\infty &= \|(I-K)^{-1}\tilde{z}\|_\infty \leq \frac{Ne^{\nu r}}{1-\|K\|}\|\nu_1 - \nu_2\| \\ &\leq \frac{Ne^{\nu r}\|\nu_1 - \nu_2\|}{1 - \frac{N(1+H)e^{\nu r}}{1-e^{-(\nu-\mu)}}(N_1\|\Lambda_1 T_1^+\varphi\|_\infty + N_2\|\Lambda_1\varphi\|_\infty)} := C_\mu\|\nu_1 - \nu_2\|.\end{aligned}$$

This yields that $w(t) \le C_\mu \|\nu_1 - \nu_2\|$ for $t \ge s$. Hence,

$$h(t) = \|u_t - v_t\|_{\mathcal{C}} \le g(t) = e^{-\mu(t-s)} w(t) \le C_\mu e^{-\mu(t-s)} \|\nu_1 - \nu_2\| \quad \text{for } t \ge s.$$

□

We now prove our main result of this section.

Theorem 2.7. *Under the hypotheses of Lemma 2.4 we let $\varphi \in E'$ be an exponentially E-invariant function and $h_\nu(\cdot)$ be defined as in Hypothesis 1.3, and define the functions $e_\nu(t) = e^{-\nu t}$ for $t \ge 0$.*
Then, if the function f is φ-Lipschitz with φ satisfying

$$\max\{\, N(1+H)e^{\nu r}(N_1 \|\Lambda_1 T_1^+ \varphi\|_\infty + N_2 \|\Lambda_1 \varphi\|_\infty),$$
$$N(1+H)e^{\nu r}(\|h_\nu\|_E + N N_1 \|e_\nu\|_E \|\varphi\|_{E'})\,\} < 1,$$

then there exists an invariant local manifold S of $\mathcal{E}$-class for the solutions to Equation (2.2).

Proof. Since $(U(t,s))_{t\ge s\ge 0}$ has an exponential dichotomy, we have that for each $t \ge 0$ the phase space $\mathcal{C}$ splits into the direct sum $\mathcal{C} = \mathrm{Im}\widetilde{P}(t) \oplus \mathrm{Ker}\widetilde{P}(t)$ where the projections $\widetilde{P}(t)$, $t \ge 0$, are defined as in Equality (2.1). Clearly, $\sup_{t\ge 0} \|\widetilde{P}(t)\| < \infty$. We now construct a local manifold $S = \{(t, S_t)\}_{t\ge 0}$ for the solutions to Eq. (2.2). To do this, we determine the surface S_t for $t \ge 0$ by the formula

$$S_t := \{\phi + \Phi_t(\phi) : \phi \in \mathbb{B}_l \cap \mathrm{Im}\widetilde{P}(t)\} \subset \mathcal{C}$$

where $l = \frac{\rho}{2NN_1 e^{\nu r}\|e_\nu\|}$, $\mathbb{B}_l := \{\phi \in \mathcal{C} : \|\phi\|_{\mathcal{C}} \le l\}$, and the operator Φ_t is defined for each $t \ge 0$ by

$$\Phi_t(\phi)(\theta) = \int_t^\infty \mathcal{G}(t-\theta, \tau) f(\tau, u_\tau) d\tau \text{ for all } \theta \in [-r, 0],$$

here $u(\cdot)$ is the unique solution of Eq. (2.2) on $[-r+t, \infty)$ satisfying the facts that $\widetilde{P}(t)u_t = \phi$ and $\chi_{[t,\infty}u_t \in \mathcal{B}_\rho \subset \mathcal{E}(\mathbb{R}_+, \mathcal{C})$ (Note that the existence and uniqueness of $u(\cdot)$ is guaranteed by Theorem 2.6). On the other hand, by the definition of the Green's function $\mathcal{G}$ (see Eq. (2.4)) we have that $\Phi_t(\phi) \in \mathrm{Ker}\widetilde{P}(t)$. We next show that the stable manifold S satisfies the conditions of Definition 2.3.

Firstly, we prove that Φ_{t_0} is local Lipschitz continuity with Lipschitz constant independent of t_0. Indeed, for $\phi \in \mathbb{B}_l \cap \mathrm{Im}\widetilde{P}(t_0)$ we have

$$\begin{aligned}\|\Phi_{t_0}(\phi)(\theta)\| &\leq N(1+H)\int_{t_0}^{\infty} e^{-\nu|t_0-\theta-\tau|}\varphi(\tau)\|u_\tau\|_C\, d\tau \\ &\leq N(1+H)e^{\nu r}\int_{t_0}^{\infty} e^{-\nu|t_0-\tau|}\varphi(\tau)\|u_\tau\|_C\, d\tau \\ &\leq N(1+H)e^{\nu r}\|h_\nu\|_E\|u\|_{\mathcal{E}} \leq \frac{\rho}{2} \text{ for all } \theta \in [-r,0] \quad (2.10)\end{aligned}$$

Therefore $\Phi_{t_0}(\phi)(\cdot) \in \mathcal{B}_{\rho/2} \subset \mathcal{B}_\rho$. Moreover, for ϕ_1 and ϕ_2 belonging to $\mathbb{B}_l \cap \mathrm{Im}\widetilde{P}(t_0)$ we have

$$\begin{aligned}\|\Phi_{t_0}(\phi_1)(\theta) - \Phi_{t_0}(\phi_2)(\theta)\| &\leq N(1+H)\int_{t_0}^{\infty} e^{-\nu|t_0-\theta-\tau|}\varphi(\tau)\|u_\tau - v_\tau\|_C\, d\tau \\ &\leq N(1+H)\int_{t_0}^{\infty} \varphi(\tau)\|u_\tau - v_\tau\|_C\, d\tau \\ &\leq N(1+H)\|\varphi\|_{E'}\|u-v\|_{\mathcal{E}}. \quad (2.11)\end{aligned}$$

Furthermore, by the Lyapunov-Perron equation for $u(\cdot)$ and $v(\cdot)$ (see Eq. (2.5)) we have

$$\|u-v\|_{\mathcal{E}} \leq Ne^{\nu r}e^{-\nu(t-t_0)}\|\phi_1-\phi_2\|_C + N(1+H)e^{\nu r}h_\nu(t)\|u-v\|_{\mathcal{E}} \text{ for } t \geq t_0.$$

By the Banach lattice property of E and $e^{-\nu(t-t_0)} = T_{t_0}^{+}e_\nu(t)$, we obtain

$$\begin{aligned}\|u-v\|_{\mathcal{E}} &\leq Ne^{\nu r}\|T_{t_0}^{+}e_\nu\|_E\|\phi_1-\phi_2\|_C + N(1+H)e^{\nu r}\|h_\nu\|_E\|u-v\|_{\mathcal{E}} \\ &\leq NN_1e^{\nu r}\|e_\nu\|_E\|\phi_1-\phi_2\|_C + N(1+H)e^{\nu r}\|h_\nu\|_E\|u-v\|_{\mathcal{E}}.\end{aligned}$$

This follows that

$$\|u-v\|_{\mathcal{E}} \leq \frac{NN_1e^{\nu r}\|e_\nu\|_E}{1-N(1+H)e^{\nu r}\|h_\nu\|_E}\|\phi_1-\phi_2\|_C.$$

Substituting this inequality to (2.10) we obtain that

$$\|\Phi_{t_0}(\phi_1)-\Phi_{t_0}(\phi_2)\|_C \leq \frac{N^2(1+H)N_1e^{\nu r}\|e_\nu\|_E\|\varphi\|_{E'}}{1-N(1+H)e^{\nu r}\|h_\nu\|_E}\|\phi_1-\phi_2\|_C.$$

Therefore, Φ_{t_0} is Lipschitz continuity with the Lipschitz constant

$$k := \frac{N^2(1+H)N_1e^{\nu r}\|e_\nu\|_E\|\varphi\|_{E'}}{1-N(1+H)e^{\nu r}\|h_\nu\|_E} < 1$$

independent of t_0.

To show that S_{t_0} is homeomorphic to $\mathbb{B}_l \cap \mathrm{Im}\widetilde{P}(t_0)$. We define the transformation $\mathbf{F} : \mathbb{B}_l \cap \mathrm{Im}\widetilde{P}(t_0) \to S_{t_0}$ by $\mathbf{F}\phi := \phi + \Phi_{t_0}(\phi)$ for all $\phi \in \mathbb{B}_l \cap \mathrm{Im}\widetilde{P}(t_0)$. Obviously, $\mathbf{F}$ is continuous and surjective. If $\mathbf{F}\phi_1 = \mathbf{F}\phi_2$ then $\phi_1 - \phi_2 = \Phi_{t_0}(\phi_2) - \Phi_{t_0}(\phi_1)$. Since Φ_{t_0} is Lipschitz mapping with Lipschitz constant $k < 1$ so $\phi_1 = \phi_2$. Thus, $\mathbf{F}$ is a bijective mapping. We also have

$$\|\phi_1 - \phi_2\|_{\mathcal{C}} \leq \|\mathbf{F}\phi_1 - \mathbf{F}\phi_2\|_{\mathcal{C}} + \|\Phi_{t_0}(\phi_2) - \Phi_{t_0}(\phi_1)\|_{\mathcal{C}} \leq \|\mathbf{F}\phi_1 - \mathbf{F}\phi_2\|_{\mathcal{C}} + k\|\phi_1 - \phi_2\|_{\mathcal{C}}.$$

Therefore, $\|\phi_1 - \phi_2\|_{\mathcal{C}} \leq \frac{1}{1-k}\|\mathbf{F}\phi_1 - \mathbf{F}\phi_2\|_{\mathcal{C}}$. This yields continuity of inverse map $\mathbf{F}^{-1}$. So, $\mathbf{F}$ is a homeomorphism. Therefore, the condition (ii) in Definition 2.3 is satisfied.

The condition (iii) in Definition 2.3 now follows from Theorem 2.6. We now prove that the condition (iv) of Definition 2.3 is satisfied. Indeed, let $u(\cdot)$ be solution of Eq. (2.2) such that the function $u_s(\theta) \in S_s$ for all $\theta \in [-r, 0]$. Then, by Lemma 2.4, the solution $u(t)$ for $t \in [s, \infty)$ can be rewritten in the form

$$u(t) = U(t, s)\nu_0 + \int_s^\infty \mathcal{G}(t, \tau) f(\tau, u_\tau) d\tau \text{ for some } \nu_0 \in \mathrm{Im}P(s).$$

Thus, for $t \geq s$ and $\theta \in [-r, 0]$ we have

$$\begin{aligned}
u(t-\theta) &= U(t-\theta, s)\nu_0 + \int_s^\infty \mathcal{G}(t-\theta, \tau) f(\tau, u_\tau) d\tau \\
&= U(t-\theta, s)\nu_0 + \int_s^t \mathcal{G}(t-\theta, \tau) f(\tau, u_\tau) d\tau + \int_t^\infty \mathcal{G}(t-\theta, \tau) f(\tau, u_\tau) d\tau \\
&= U(t-\theta, s)\nu_0 + \int_s^t U(t-\theta, \tau) P(\tau) f(\tau, u_\tau) d\tau + \int_t^\infty \mathcal{G}(t-\theta, \tau) f(\tau, u_\tau) d\tau \\
&= U(t-\theta, t)\left[U(t, s)\nu_0 + \int_s^t U(t, \tau) P(\tau) f(\tau, u_\tau) d\tau\right] + \int_t^\infty \mathcal{G}(t-\theta, \tau) f(\tau, u_\tau) d\tau
\end{aligned}$$

Put $\mu_0 = U(t, s)\nu_0 + \int_s^t U(t, \tau) P(\tau) f(\tau, u_\tau) d\tau$. We have $P(t)\mu_0 = \mu_0$, hence $\mu_0 \in \mathrm{Im}P(t)$. We thus obtain that $U(t-\theta, t)\mu_0$ belongs to $\mathrm{Im}\widetilde{P}(t)$ and

$$u(t-\theta) = U(t-\theta, t)\mu_0 + \int_t^\infty \mathcal{G}(t-\theta, \tau) f(\tau, u_\tau) d\tau.$$

By the uniqueness of $u(\cdot)$ on $[s-r, \infty)$ as in the proof of Theorem 2.6 we have that Eq. (2.2) has a unique solution $u(\cdot)$ on $[-r+t, \infty)$ satisfying $(\widetilde{P}(t)u_t)(\theta) = U(t-\theta, t)\mu_0$ and

$$u(\xi) = U(2t-\xi, t)\mu_0 + \int_t^\infty \mathcal{G}(2t-\xi, \tau) f(\tau, u_\tau) d\tau$$

for $\xi \in [-r+t, t]$. Therefore, the history function u_t can be viewed as

$$u_t(\theta) = u(t+\theta) = U(t-\theta, t)\mu_0 + \int_t^{\infty} \mathcal{G}(t-\theta, \tau) f(\tau, u_\tau) d\tau = \phi(\theta) + \Phi_t(\phi)(\theta)$$

$$\text{for all } \theta \in [-r, 0].$$

Hence, $u_t \in S_t$ for $t \geq s$. □

3. HUTCHINSON MODELS

In this final section, we give an illustrating example for the attained results. Concretely, we consider the modified Hutchinson equation with non-autonomous diffusion

$$\begin{cases} \dfrac{\partial w(x,t)}{\partial t} = a(t)[\dfrac{\partial^2 w(x,t)}{\partial x^2} + \eta w(x,t)] + \psi(t) w(x, t-1)(1 + w(x,t)) \text{ for } 0 < x < \pi,\ t \geq 0, \\ \dfrac{\partial w(0,t)}{\partial x} = \dfrac{\partial w(\pi,t)}{\partial x} = 0, \quad t \geq 0, \\ w(x,\theta) = \phi(x,\theta) \in \mathcal{C} \quad \text{for } 0 < x < \pi,\ \theta \in [-1, 0], \end{cases} \tag{3.1}$$

here, $\eta \in \mathbb{R}$ and $\eta \neq n^2$ for all $n \in \mathbb{N}$; the function $a(\cdot) \in L_{1,loc}(\mathbb{R}_+)$ satisfies the condition $0 < \gamma_0 \leq a(t) \leq \gamma_1$ for all $t \in \mathbb{R}_+$ and fixed constants γ_0, γ_1.

We next put $X := C^2[0, \pi]$, $\mathcal{C} := C([-1, 0], X)$, and let $A : X \supset D(A) \to X$ be defined by $Ay = y'' + \eta y$, with the domain

$$D(A) = \{y \in X : y'(0) = y'(\pi) = 0\}.$$

It can be seen (see [6]) that A is the generator of an analytic semigroup $(\mathbb{T}(t))_{t\geq 0}$. Since $\sigma(A) = \{-n^2 + \eta : n = 1, 2, 3, \cdots\}$ applying the spectral mapping theorem for analytic semigroups we get

$$\sigma(\mathbb{T}(t)) = e^{t\sigma(A)} = \{e^{t(-n^2+\eta)} : n = 1, 2, 3, ...\} \tag{3.2}$$

$$\text{and hence } \sigma(\mathbb{T}(t)) \cap \Gamma = \emptyset \text{ for all } t > 0,$$

where $\Gamma := \{\lambda \in \mathbb{C} : |\lambda| = 1\}$.

Putting now $A(t) := a(t)A$ we have that the family $(A(t))_{t\geq 0}$ generates evolution family $U(t, s)_{t\geq s\geq 0}$ which is defined by the formula $U(t, s) = \mathbb{T}(\int_s^t a(\tau)d\tau)$.

By (3.2) we have that the analytic semigroup $(\mathbb{T}(t))_{t\geq 0}$ is hyperbolic (or has an exponential dichotomy) with the projection P satisfying

1. $\|\mathbb{T}(t)x\| \leq Ne^{-\beta t}\|x\|$ for $x \in PX$, $t \geq 0$
2. $\|\mathbb{T}(-t)_{|}x\| = \|(\mathbb{T}(t)_{|})^{-1}x\| \leq Ne^{-\beta t}\|x\|$ for $x \in \text{Ker}P$, $t \geq 0$, where the invertible operator $\mathbb{T}(t)_{|}$ is the restriction of $T(t)$ to Ker$P$, and N, β are positive constants.

Using the hyperbolicity of $(\mathbb{T}(t))_{t\geq 0}$ it is straightforward to check that the evolution family $U(t,s)_{t\geq s\geq 0}$ has an exponential dichotomy with the projection $P(t) = P$ for all $t \geq 0$ and the dichotomy constants N and $\nu := \beta\gamma_0$ by the following estimates:

$$\|U(t,s)x\| \leq Ne^{-\nu(t-s)}\|x\| \text{ for } x \in PX,\ t \geq s \geq 0,$$
$$\|U(s,t)_{|}x\| \leq Ne^{-\nu(t-s)}\|x\| \text{ for } x \in \text{Ker}P,\ t \geq s \geq 0.$$

We then define the function and $f : \mathbb{R}_+ \times \mathcal{C} \to X$ by $f(t,\varphi) := \psi(t)(\delta_{-1}\varphi)(1+\varphi)$ for $\varphi \in \mathcal{C} := C([-1,0],X)$ where δ_{-1} is the Dirac delta function concentrated at -1,

$$\begin{cases} \dfrac{d}{dt}u(\cdot,t) = A(t)u(\cdot,t) + f(t,u_t(\cdot,\theta)), \\ u_0(\cdot,\theta) = \phi(\cdot,\theta) \in \mathcal{C}. \end{cases}$$

Moreover,

$$\begin{aligned} \|f(t,u_t(\cdot,\theta)) - f(t,v_t(\cdot,\theta))\| &= \psi(t)\|u_t(\cdot,\theta)(1+u(\cdot,t)) - v_t(\cdot,\theta)(1+v(\cdot,t))\| \\ &= \psi(t)\|u_t(\cdot,\theta) - v_t(\cdot,\theta) + u_t(\cdot,\theta)u(\cdot,t) - v_t(\cdot,\theta)u(\cdot,t) + \\ &\quad + v_t(\cdot,\theta)u(\cdot,t) - v_t(\cdot,\theta)v(\cdot,t)\| \\ &\leq \psi(t)(\|u_t(\cdot,\theta) - v_t(\cdot,\theta)\| + \|u_t(\cdot,\theta)u(\cdot,t) - v_t(\cdot,\theta)u(\cdot,t)\| + \\ &\quad + \|v_t(\cdot,\theta)u(\cdot,t) - v_t(\cdot,\theta)v(\cdot,t))\| \\ &\leq (1+2\rho)\psi(t) \sup_{\theta\in[-1,0]} \|u_t - v_t\| \quad \text{for all } u_t, v_t \in \mathfrak{B}_\rho. \end{aligned}$$

We now choose $\omega(t) := be^{-\alpha t}$ for $t \in \mathbb{R}_+$ with constants $\alpha > \nu$, and $b > 0$. Putting $\varphi(t) = (1+2\rho)\omega(t)$ for $t \in \mathbb{R}_+$ we can see that $\varphi \in L_p(\mathbb{R}_+)$ for $1 \leq p < +\infty$ and

$$\|\varphi\|_{L_p} = (1+2\rho)b\left(\int_0^\infty e^{-p\alpha t}dt\right)^{\frac{1}{p}} = (1+2\rho)b\left(\frac{1}{\alpha p}\right)^{\frac{1}{p}}.$$

Then, we prove that the function

$$h_\nu(t) = \left(\int_0^\infty e^{-p\nu|t-\tau|}\varphi^p(\tau)d\tau\right)^{\frac{1}{p}} \quad \text{for } t \in \mathbb{R}_+$$

belongs L_q for q satisfying $\frac{1}{p}+\frac{1}{q}=1$. Indeed, for $t\geq 0$ we compute

$$\begin{aligned} h_\nu(t) &= \left(\int_0^\infty e^{-p\nu|t-\tau|}\varphi^p(\tau)d\tau\right)^{\frac{1}{p}} = \left(\int_0^\infty e^{-p\nu|t-\tau|}((1+2\rho)be^{-\alpha\tau})^p d\tau\right)^{\frac{1}{p}} \\ &= (1+2\rho)b\left(\int_0^t e^{-p\nu|t-\tau|}e^{-\alpha p\tau}d\tau + \int_t^{+\infty} e^{-p\nu|t-\tau|}e^{-\alpha p\tau}d\tau\right)^{\frac{1}{p}} \\ &= (1+2\rho)b\left(\frac{e^{-p\nu t}-e^{-p\alpha t}}{p(\alpha-\nu)} + \frac{e^{-p\alpha t}}{p(\alpha+\nu)}\right)^{\frac{1}{p}}. \end{aligned}$$

Therefore, $h_\nu \in L_q(\mathbb{R}_+)$ for $\frac{1}{q}+\frac{1}{p}=1$ and

$$\begin{aligned} \|h_\nu\|_{L_q} &= \left(\int_0^\infty (h_\nu(t))^q\,dt\right)^{\frac{1}{q}} \\ &\leq (1+2\rho)b\left(\frac{2\alpha}{p(\alpha+\nu)(\alpha-\nu)}\right)^{\frac{1}{p}}\left(\frac{1}{\nu q}\right)^{\frac{1}{q}}. \end{aligned} \tag{3.3}$$

By Theorem 2.7 we see that if

$$(1+2\rho)bN(1+H)e^{\nu r}\left(\frac{1}{\nu q}\right)^{\frac{1}{q}}\left[NN_1\left(\frac{1}{\alpha p}\right)^{\frac{1}{p}} + \left(\frac{2\alpha}{p(\alpha+\nu)(\alpha-\nu)}\right)^{\frac{1}{p}}\right] < 1$$

and

$$(1+2\rho)bN(1+H)e^{\nu r}\left(\frac{1}{\nu q}\right)^{\frac{1}{q}}\left(\frac{2\alpha}{p(\alpha+\nu)(\alpha-\nu)}\right)^{\frac{1}{p}} < \frac{1}{2}$$

then there is an local-stable manifold of L_p-class for mild solutions to Equation (3.1).

ACKNOWLEDGMENT

This research is funded by Vietnam National Foundation for Science and Technology Development (NAFOSTED).

REFERENCES

[1] Bates P., Lu K., Zeng C., Existence and persistence of invariant manifolds for semiflows in Banach space, *Mem. Amer. Math. Soc.* 135 (1998), 645.

[2] Bates P., Lu K., Zeng C., Invariant foliations near normally hyperbolic invariant manifolds for semiflows, *Trans. Amer. Math. Soc.* 352 (10) (2000), 4641-4676.

[3] Bates P., Jones C., Invariant manifolds for semilinear partial differential equations, *Dyn. Rep.* 2 (1989), 1 - 38.

[4] Carr J., Applications of Centre Manifold Theory, *Applied Mathematical Sciences* 35, Springer-Verlag, New York-Berlin, 1981.

[5] Chicone C., Latushkin Y., Center manifolds for infinite dimensional nonautonomous differential equations, *J. Differential Equations* 141 (1997), 356-399.

[6] Engel K. J., Nagel R., One-parameter Semigroups for Linear Evolution Equations, *Graduate Text Math.* 194, Springer-Verlag, Berlin-Heidelberg, 2000.

[7] Hale J. K., Flows on center manifolds for scalar functional differential equations, *Proc. Roy. Soc. Edinburgh A* 101 (1985), 193-201.

[8] Hirsch N, Pugh C., Shub M., *Invariant Manifolds, Lecture Notes in Math* vol. 183, Springer, New York, 1977.

[9] Minh N. V, Wu J., Invariant manifolds of partial functional differential equations, *J. Differential Equations* 198 (2004), 381 - 421.

[10] Latushkin Y, Prüss J., Schnaubelt R., Center manifolds and dynamics near equilibria of quasilinear parabolic systems with fully nonlinear boundary conditions, *Disc. Cont. Dyn. Sys.* -B 9 (2008), 595-633.

[11] Murray J. D., *Mathematical Biology I: An Introduction*, Springer-Verlag Berlin, 2002.

[12] Murray J. D., *Mathematical Biology II: Spatial Models and Biomedical Applications*, Springer-Verlag Berlin, 2003.

[13] Yagi A., *Abstract Parabolic Evolution Equations and their Applications*, Springer Verlag, 2009.

[14] Huy N. T., Invariant manifolds of admissible classes for semi-linear evolution equations, *J. Differential Equations* 246 (2009), 1820-1844.

[15] Huy N. T., Ha V. T. N., Admissibly Integral Manifolds for Semi-linear Evolution Equations, *Annal. Pol. Math.*, 112 (2014), 127-163.

[16] Huy N. T., Stable manifolds for semi-linear evolution equations and admissibility of function spaces on a half-line, *J. Math. Anal. Appl.* 354 (2009), 372 - 386.

[17] Huy N.T., Exponential dichotomy of evolution equations and admissibility of function spaces on a half-line, *J. Funct. Anal.* 235 (2006), 330 - 354.

[18] Huy N. T., Duoc T. V., Integral manifolds for partial functional differential equations in admissible spaces on a half-line, *J. Math. Anal. Appl.*, 411 (2014), 816-828.

[19] Huy N. T., Duoc T. V., Khanh D. X., Attraction Property of Admissible Integral Manifolds and Applications to Fisher-Kolmogorov Model, *Taiwanese J. Math.* 20 (2016), 365-385.

[20] Massera J. J., Schäffer J. J., *Linear Differential Equations and Function Spaces*, Academic Press, New York, 1966.

[21] Minh N. V., Räbiger F., Schnaubelt R., Exponential stability, exponential expansiveness and exponential dichotomy of evolution equations on the half line, *Integral Equations Operator Theory* 32 (1998), 332 - 353.

[22] Nagel R., Nickel G., Well-posedness for non-autonomous abstract Cauchy problems, *Progr. Nonlinear Differential Equations Appl.* 50 (2002), 279 - 293.

[23] Pazy A., *Semigroup of Linear Operators and Application to Partial Differential Equations*, Springer-Verlag, Berlin, 1983.

[24] Räbiger F., Schnaubelt R., The spectral mapping theorem for evolution semigroups on spaces of vector-valued functions, *Semigroup Forum* 48 (1996), 225 - 239.

[25] Schnaubelt R., Asymptotically autonomous parabolic evolution equations, *J. Evol. Equ.* 1 (2001), 19-37.

[26] Wu J., *Theory and Applications of Partial Functional Differential Equations*, Springer Verlag, 1996.

In: Evolutionary Processes and Applications ISBN: 978-1-53616-350-6
Editor: G. M. N'Guérékata

Chapter 8

STACKELBERG CONTROL OF A BACKWARD LINEAR HEAT EQUATION

***L. L. Djomegne Njoukoue*[1,*], *G. Mophou*[2,†] *and G. Deugoue*[1,‡]**
[1]University of Dschang, Dschang, Cameroon, West Region
[2]African Institute for Mathematical Sciences (AIMS), Limbe Crystal Gardens, South West Region, Cameroon

Abstract

We are interested in the hierarchic control problem for a linear backward heat equation. We assume that we can act on the system via two distributed controls. One control called follower solves an optimal control problem which consists in bringing the flux of the system to zero. The other control called leader solves a null controllability problem. The results are achieved by means of inequality of observability associated to a non-homogeneous Dirichlet boundary condition.

Keywords: optimal control, Carleman inequality, Stackelberg control, heat equation

*Corresponding Author's Email: landry.djomegne@yahoo.fr.
†Corresponding Author's Email: gisele.mophou@univ-antilles.fr.
‡Corresponding Author's Email: agdeugoue@yahoo.fr.

1. INTRODUCTION

Let $n \in \mathbb{N} \setminus \{0\}$ and Ω be a bounded open subset of $\mathbb{R}^n$ with boundary Γ of class $\mathcal{C}^2$. Let Γ_1 be a subset of Γ. Let also $\mathcal{O}$, ω_1 and ω_2 be three non-empty subsets of Ω such that $\omega_1 \subset \omega_2$. For a time $T > 0$, we set $Q = (0,T) \times \Omega$, $\mathcal{O}_T = (0,T) \times \mathcal{O}$, $\omega_1^T = (0,T) \times \omega_1$, $\omega_2^T = (0,T) \times \omega_2$, $\Sigma = (0,T) \times \Gamma$ and $\Sigma_1 = (0,T) \times \Gamma_1$. Then, we consider the following linear heat equation:

$$\left\{ \begin{array}{rcll} -\dfrac{\partial q}{\partial t} - \Delta q + a_0 q & = & h_0 \mathbf{1}_{\mathcal{O}} + v\mathbf{1}_{\omega_1} + k\mathbf{1}_{\omega_2} & \text{in} \quad Q, \\ q & = & 0 & \text{on} \quad \Sigma, \\ q(T) & = & 0 & \text{in} \quad \Omega, \end{array} \right. \tag{1.1}$$

where $a_0 \in L^\infty(Q)$, $h_0 \in L^2(\mathcal{O}_T)$, the control v and k belong respectively to $L^2(\omega_1^T)$ and $L^2(\omega_2^T)$. The function $\mathbf{1}_{\mathbb{X}}$ denotes the characteristic function on the open set $\mathbb{X}$. Under the above assumptions on the data, it is well known that system (1.1) has a unique solution $q(v,k) := q(t,x;v,k) \in H^{2,1}(Q) \cap L^2((0,T); H_0^1(\Omega))$ where from now on,

$$H^{r,s}((0,T) \times \mathbb{X}) = L^2((0,T); H^r(\mathbb{X})) \cap H^s((0,T); L^2(\mathbb{X})).$$

Moreover, there exists a positive constant $C = C(\|a_0\|_{L^\infty(Q)}, T)$ such that

$$\|q\|_{H^{2,1}(Q)} \leq C \left(\|h_0\|_{L^2(\mathcal{O}_T)} + \|v\|_{L^2(\omega_1^T)} + \|k\|_{L^2(\omega_2^T)} \right), \tag{1.2}$$

and it follows from the continuity of the trace that,

$$\left\| \frac{\partial q}{\partial \nu}(v,k) \right\|_{H^{\frac{1}{2},\frac{1}{4}}(\Sigma)} \leq C \left(\|h_0\|_{L^2(\mathcal{O}_T)} + \|v\|_{L^2(\omega_1^T)} + \|k\|_{L^2(\omega_2^T)} \right). \tag{1.3}$$

We refer to [1] and the reference therein for more literature on the regularity results on the parabolic equations.

In this chapter we are concerned with the following problems:

Problem 1. *(Optimal control problem)*

Let ω_1 and ω_2 be two non-empty subsets of Ω. Given $k \in L^2(\omega_2^T)$, find the control $u := u(k) \in L^2(\omega_1^T)$ such that

$$J(u) = \inf_{v \in L^2(\omega_1^T)} J(v) \tag{1.4}$$

with

$$J(v) = \left\| \frac{\partial q}{\partial \nu}(v,k) \right\|^2_{L^2(\Sigma_1)} + N\|v\|^2_{L^2(\omega_1^T)}, \tag{1.5}$$

where $N > 0$, ν *is the unit exterior normal vector of* Γ, $\frac{\partial q}{\partial \nu}$ *is the normal derivative of* q *with respect to* ν *and* $q(v,k)$ *is solution of system* (1.1).

Problem 2. *(Null controllability problem)*

Let ω_1 *and* ω_2 *be two non-empty subsets of* Ω *such that* $\omega_1 \subset \omega_2$. *Let also* $u(k)$ *be the optimal control obtained in the Problem 1. Then, find a control* $k \in L^2(\omega_2^T)$ *such that if* $q = q(t,x;u(k),k)$ *is solution of (1.1), then*

$$q(0) = q(0,x;u(k),k) = 0 \text{ for } x \in \Omega. \tag{1.6}$$

Remark 1. *Note that from* (1.3), *the cost function defined by* (1.5) *is well defined.*

Problem 1 and Problem 2 form a Stackelberg control problem which is a hierarchical optimization problem with k acting as the "leader" and v, the "follower".

The Stackelberg leadership model [2] is a strategic game in economics in which two firms compete on the market of the same product. The first to act must integrate the reaction of the other company in the choices it makes in the amount of product that it decides to put on the market. There are few works in the literature on Stackelberg's control of partial differential equations. J.L. Lions in [3] used this concept for a linear parabolic equation with two controls; the one called follower aimed to bring the state of the system to a desired state, while the other control the leader solved a problem of approximate controllability. O. Nakoulima [4] used the concept of Stackelberg for a linear and backward parabolic equation with two controls to be determined. In his work, the follower solved a null controllability problem with constraints on control while the leader solved an optimal control problem. In [5, 6], M. Mercan revisited the notion of controllability in the sense of Stackelberg given by O. Nakoulima [4] by choosing the follower of minimal norm. This new notion is then applied by M. Mercan and O. Nakoulima in [7] on the controllability of a two-stroke problem with constraint on the states. The results were obtained by means of Carleman inequality adapted to the constraints. Recently F. D Araruna et *al.* [8] studied the Stackelberg-Nash strategies to the control of parabolic equations. This strategy involves three controls: one leader and two followers. The leader is assumed to choose the policy and a Nash equilibrium pair governs the action of the followers. In [9] L. Teresa et *al.* used the Stackelberg control to combine the concepts of controllability with robustness. In their work the leader is responsible for a controllability to trajectories objective and the follower solves a robust control problem.

Motivated by all the above works, we study the Stackelberg control for a linear backward parabolic equation where the follower solves an optimal control problem with boundary observation and the leader a null controllability problem. The novelty of the work lies on the fact that the null controllability is achieved by means of Carleman inequality associated to a non-homogeneous Dirichlet boundary condition. More precisely, we prove the following results:

Theorem 2. *Let $\Omega \subset \mathbb{R}^n$, $n \in \mathbb{N} \setminus \{0\}$ be a bounded subset with boundary Γ of class C^2. Let Γ_1 be a subset of Γ. Let ω_1 and ω_2 be two non empty subsets of Ω. Let also $k \in L^2(\omega_2^T)$. Then, there exists $p \in L^2(Q)$ such that the optimal control problem* (1.4) *has a unique solution $u \in L^2(\omega_1^T)$ which is characterized by the following optimality system:*

$$\begin{cases} -\dfrac{\partial q}{\partial t} - \Delta q + a_0 q &= h_0 \boldsymbol{1}_{\mathcal{O}} + u\boldsymbol{1}_{\omega_1} + k\boldsymbol{1}_{\omega_2} \quad in \quad Q, \\ q &= 0 \quad on \quad \Sigma, \\ q(T) &= 0 \quad in \quad \Omega, \end{cases} \tag{1.7}$$

$$\begin{cases} \dfrac{\partial p}{\partial t} - \Delta p + a_0 p &= 0 \quad in \quad Q, \\ p &= \frac{\partial q}{\partial \nu}\boldsymbol{1}_{\Gamma_1} \quad on \quad \Sigma, \\ p(0) &= 0 \quad in \quad \Omega, \end{cases} \tag{1.8}$$

and

$$u = \frac{p}{N} \ in \ \omega_1^T. \tag{1.9}$$

Moreover there exists a constant $C = C(\|a\|_{L^\infty(Q)}, T, N) > 0$ such that

$$\|u\|_{L^2(\omega_1^T)} \le C\left(\|h_0\|_{L^2(\mathcal{O}_T)} + \|k\|_{L^2(\omega_2^T)}\right). \tag{1.10}$$

Theorem 3. *Assume that the assumptions of Theorem 2 hold. Assume also that ω_1 and ω_2 are such that $\omega_1 \subset \omega_2$. Then, there exists a positive weight function θ (a precise definition of θ will be given later on) such that for any $h_0 \in L^2(\mathcal{O}_T)$ with $\theta h_0 \in L^2(\mathcal{O}_T)$, there exists a unique control $\hat{k} \in L^2(\omega_2^T$*

such that $(\hat{k}, q, p)$ *is solution of the null controllability problem* (1.7)-(1.9) *and* (1.6). *Moreover*

$$\hat{k} = \hat{\rho} \text{ in } \omega_2^T \tag{1.11}$$

where $\hat{\rho}$ *satisfies*

$$\begin{cases} \dfrac{\partial \hat{\rho}}{\partial t} - \Delta \hat{\rho} + a_0 \hat{\rho} &= 0 & \text{in} & Q, \\ \hat{\rho} &= \frac{\partial \hat{\Psi}}{\partial \nu} \boldsymbol{1}_{\Gamma_1} & \text{on} & \Sigma, \end{cases} \tag{1.12}$$

with $\hat{\Psi}$ *solution of*

$$\begin{cases} -\dfrac{\partial \hat{\Psi}}{\partial t} - \Delta \hat{\Psi} + a_0 \hat{\Psi} &= \frac{\hat{\rho}}{N} \boldsymbol{1}_{\omega_1} & \text{in} & Q, \\ \hat{\Psi} &= 0 & \text{on} & \Sigma, \\ \hat{\Psi}(T) &= 0 & \text{in} & \Omega. \end{cases} \tag{1.13}$$

In addition, there exists a constant $C = C(C_1, \|a_0\|_{L^\infty(Q)}, T, N)$ *such that*

$$\|\hat{k}\|_{L^2(\omega_2^T)} \leq \sqrt{C} \|\theta h_0\|_{L^2(\mathcal{O}_T)}. \tag{1.14}$$

The rest of this chapter is organized as follows: In Section 2 we prove the Theorem 2. In Section 3, we establish an appropriate inequality of Carleman useful to solve Problem 2. Section 4 is devoted to the proof of the Theorem 3. A conclusion is given in Section 4.

From now on, we will denote by $C(X)$ a generic positive constant whose value varies from a line to another but depending on X.

2. Proof of Theorem 2

In this section, we are interested in the resolution of Problem 1.

Proposition 1. *For any* $k \in L^2(\omega_2^T)$, *there exists a unique optimal control* $u := u(k) \in L^2(\omega_1^T)$ *such that* (1.4) *holds true.*

Proof. Since $J(v) \geq 0$ for all $v \in L^2(\omega_1^T)$, we deduce that $\inf_{v \in L^2(\omega_1^T)} J(v)$ exists. Let $(v_n) \subset L^2(\omega_1^T)$ be a minimizing sequence such that

$$\lim_{n \to +\infty} J(v_n) = \inf_{v \in L^2(\omega_1^T)} J(v). \tag{2.1}$$

Then, $q_n := q(v_n)$ is solution of system (1.1). This means that q_n satisfies

$$\left\{\begin{array}{rcll} -\dfrac{\partial q_n}{\partial t} - \Delta q_n + a_0 q_n & = & h_0 \mathbf{1}_{\mathcal{O}} + v_n \mathbf{1}_{\omega_1} + k \mathbf{1}_{\omega_2} & \text{in} \quad Q, \\ q_n & = & 0 & \text{on} \quad \Sigma, \\ q_n(T) & = & 0 & \text{in} \quad \Omega. \end{array}\right. \tag{2.2}$$

Moreover, in view of (2.1), there exists a constant $C > 0$ independent of n such that

$$J(v_n) \leq C. \tag{2.3}$$

Therefore, from the structure of J we get

$$\|v_n\|_{L^2(\omega_1^T)} \leq C(N), \tag{2.4}$$

$$\left\|\frac{\partial q_n}{\partial \nu}\right\|_{L^2(\Sigma_1)} \leq C. \tag{2.5}$$

In view of (2.2),(2.4) and (1.2), we deduce that

$$\|q_n\|_{H^{2,1}(Q)} \leq C\left(\|h_0\|_{L^2(\mathcal{O}_T)} + \|k\|_{L^2(\omega_2^T)}\right), \tag{2.6}$$

where $C = C(\|a_0\|_{L^\infty(Q)}, T, N) > 0$. Hence, there exist $u \in L^2(\omega_1^T)$, $\beta \in L^2(\Sigma_1)$, $q \in H^{2,1}(Q)$ and subsequences extracted from (v_n), $(\dfrac{\partial q_n}{\partial \nu})$ and (q_n) (still called (v_n), $(\frac{\partial q_n}{\partial \nu})$ and (q_n)) such that

$$v_n \rightharpoonup u \text{ weakly in } L^2(\omega_1^T), \tag{2.7a}$$

$$q_n \rightharpoonup q \text{ weakly in } H^{2,1}(Q), \tag{2.7b}$$

$$\frac{\partial q_n}{\partial \nu} \rightharpoonup \beta \text{ weakly in } L^2(\Sigma_1). \tag{2.7c}$$

Let $\mathbb{D}(Q)$ the set of functions of class $\mathcal{C}^\infty$ on Q with compact support. If we multiply the first equation of (2.2) by $\varphi \in \mathbb{D}(Q)$ and integrate by parts over Q, we obtain

$$\int_Q q_n \left(\frac{\partial \varphi}{\partial t} - \Delta\varphi + a_0\varphi\right) dxdt = \int_Q (h_0 \mathbf{1}_{\mathcal{O}} + v_n \mathbf{1}_{\omega_1} + k\mathbf{1}_{\omega_2})\varphi \, dxdt,$$

$$\forall \varphi \in \mathbb{D}(Q). \tag{2.8}$$

Passing to the limit in (2.8) while using (2.7a) and (2.7b), we obtain

$$\int_Q q\left(\frac{\partial\varphi}{\partial t}-\Delta\varphi+a_0\varphi\right)dxdt=\int_Q(h_0\mathbf{1}_{\mathcal{O}}+u\mathbf{1}_{\omega_1}+k\mathbf{1}_{\omega_2})\varphi\,dxdt,$$
$$\forall\varphi\in\mathbb{D}(Q),$$

which after an integration by parts gives

$$\int_Q\left(-\frac{\partial q}{\partial t}-\Delta q+a_0q\right)\varphi dxdt=\int_Q(h_0\mathbf{1}_{\mathcal{O}}+u\mathbf{1}_{\omega_1}+k\mathbf{1}_{\omega_2})\varphi\,dxdt,$$
$$\forall\varphi\in\mathbb{D}(Q).$$

Thus, we obtain

$$-\frac{\partial q}{\partial t}-\Delta q+a_0q=h_0\mathbf{1}_{\mathcal{O}}+u\mathbf{1}_{\omega_1}+k\mathbf{1}_{\omega_2}\ \text{ in }\ Q. \tag{2.9}$$

As $q\in H^{2,1}(Q)$, we have the existence of the traces $q_{\,\Sigma},\dfrac{\partial q}{\partial\nu}_{|\Sigma}\in L^2(\Sigma)$ and $q(T)\in L^2(\Omega)$. So if we multiplying the first equation of (2.2) by $\varphi\in C^\infty(\overline{Q})$ such that $\varphi=0$ on $\Sigma\setminus\Sigma_1$ and $\varphi(0)=0$ in Ω and integrate by parts over Q, we obtain

$$\begin{aligned}&-\int_{\Sigma_1}\frac{\partial q_n}{\partial\nu}\varphi d\sigma dt+\int_Q q_n\left(\frac{\partial\varphi}{\partial t}-\Delta\varphi+a_0\varphi\right)dxdt\\&=\int_Q\left(h_0\mathbf{1}_{\mathcal{O}}+v_n\mathbf{1}_{\omega_1}+k\mathbf{1}_{\omega_2}\right)\varphi dxdt.\end{aligned}\tag{2.10}$$

Passing to the limit in (2.10) when $n\to+\infty$ while using (2.7a)-(2.7c), we obtain

$$\begin{aligned}&-\int_{\Sigma_1}\beta\varphi dxdt+\int_Q q\left(\frac{\partial\varphi}{\partial t}-\Delta\varphi+a_0\varphi\right)dxdt\\&=\int_Q\left(h_0\mathbf{1}_{\mathcal{O}}+u\mathbf{1}_{\omega_1}+k\mathbf{1}_{\omega_2}\right)\varphi dxdt,\\&\forall\varphi\in C^\infty(\overline{Q})\text{ with }\varphi=0\text{ on }\Sigma\setminus\Sigma_1,\ \varphi(0)=0\text{ in }\Omega.\end{aligned}\tag{2.11}$$

Integrating by parts this latter identity, we obtain

$$\begin{aligned}
&-\int_{\Sigma_1} \beta\varphi dxdt + \int_\Omega q(T)\varphi(T)dx - \int_\Sigma \frac{\partial\varphi}{\partial\nu} q d\sigma dt \\
&+ \int_{\Sigma_1} \varphi\frac{\partial q}{\partial\nu} d\sigma dt + \int_Q \left(-\frac{\partial q}{\partial t} - \Delta q + a_0 q\right)\varphi\, dxdt \\
&= \int_Q (h_0\mathbf{1}_{\mathcal{O}} + u\mathbf{1}_{\omega_1} + k\mathbf{1}_{\omega_2})\varphi dxdt, \\
&\forall\varphi \in C^\infty(\overline{Q}) \text{ with } \varphi = 0 \text{ on } \Sigma\setminus\Sigma_1,\ \varphi(0) = 0 \text{ in } \Omega,
\end{aligned} \tag{2.12}$$

which in view of (2.9) yields

$$\begin{aligned}
&\int_{\Sigma_1} (\frac{\partial q}{\partial\nu} - \beta)\varphi\, d\sigma dt + \int_\Omega q(T)\varphi(T)dx - \int_\Sigma q\frac{\partial\varphi}{\partial\nu}\, d\sigma dt = 0, \\
&\forall\varphi \in C^\infty(\overline{Q}) \text{ with } \varphi = 0 \text{ on } \Sigma\setminus\Sigma_1,\ \varphi(0) = 0 \text{ in } \Omega.
\end{aligned} \tag{2.13}$$

Choosing successively in (2.13), $\varphi(T) = 0$ in Ω and $\varphi = 0$ on Σ_1, then $\varphi(T) = 0$ in Ω, we successively get

$$q = 0 \text{ in } \Sigma \tag{2.14}$$

and

$$\beta = \frac{\partial q}{\partial\nu} \text{ on } \Sigma_1. \tag{2.15}$$

Finally, it follows from (2.13) that

$$q(T) = 0 \text{ in } \Omega. \tag{2.16}$$

From (2.7c) and (2.15), we have

$$\frac{\partial q_n}{\partial\nu} \rightharpoonup \frac{\partial q}{\partial\nu} \text{ weakly in } L^2(\Sigma_1). \tag{2.17}$$

In view of (2.9), (2.14) and (2.16), we deduce that $q = q(u(k))$ is solution of (1.1). Using the weak lower semi-continuity of the function $v \mapsto J(v)$, (2.17) and (2.7a), we obtain that

$$J(u) \le \liminf_{n\to+\infty} J(v_n),$$

which according to (2.1) implies that

$$J(u) \le \inf_{v\in L^2(\omega_1^T)} J(v).$$

Thus

$$J(u) = \inf_{v \in L^2(\omega_1^T)} J(v).$$

The uniqueness of u comes from the fact that the functional J is strictly convex.

□

In order to characterize the optimal control u, we express the Euler-Lagrange optimality conditions:

$$\lim_{\lambda \to 0} \frac{J(u + \lambda v) - J(u)}{\lambda} = 0, \quad \text{for all } v \in L^2(\omega_1^T). \tag{2.18}$$

After some calculations, (2.18) gives

$$\int_{\Sigma_1} \frac{\partial z(v)}{\partial \nu} \frac{\partial q}{\partial \nu} \, d\sigma dt + N \int_0^T \int_{\omega_1} uv \, dx dt = 0, \; \forall v \in L^2(\omega_1^T), \tag{2.19}$$

where $z = z(v) \in H^{2,1}(Q)$ is solution of

$$\left\{ \begin{array}{rcll} -\dfrac{\partial z}{\partial t} - \Delta z + a_0 z & = & v \mathbf{1}_{\omega_1} & \text{in} \quad Q, \\ z & = & 0 & \text{on} \quad \Sigma, \\ z(T) & = & 0 & \text{in} \quad \Omega. \end{array} \right. \tag{2.20}$$

To interpret (2.19), we consider the adjoint state solution of system (1.8):

$$\left\{ \begin{array}{rcll} \dfrac{\partial p}{\partial t} - \Delta p + a_0 p & = & 0 & \text{in} \quad Q, \\ p & = & \dfrac{\partial q}{\partial \nu} \mathbf{1}_{\Gamma_1} & \text{on} \quad \Sigma, \\ p(0) & = & 0 & \text{in} \quad \Omega. \end{array} \right.$$

Since $q \in H^{2,1}(Q)$, we have that $\dfrac{\partial q}{\partial \nu} \in H^{\frac{1}{2},\frac{1}{4}}(\Sigma) \subset L^2(\Sigma)$ (see Theorem 2.1, [1], Vol. 2). Therefore we can prove by transposition that there exists a unique $p \in L^2(Q)$ such that

$$\int_{\Sigma_1} \frac{\partial q}{\partial \nu} \frac{\partial \Phi}{\partial \nu} d\sigma \, dt + \int_Q p \left(-\frac{\partial \Phi}{\partial t} - \Delta \Phi + a_0 \Phi \right) dx dt = 0,$$

$$\forall \Phi \in H^{2,1}(Q) \text{ such that } \Phi|_\Sigma = 0, \; \Phi(T) = 0 \text{ in } \Omega.$$

Thus, multiplying the first equation of (2.20) by p and integrating by parts over Q, we obtain:

$$\int_{\Sigma_1} \frac{\partial z(v)}{\partial \nu}\frac{\partial q}{\partial \nu}\, d\sigma dt = -\int_0^T \int_{\omega_1} vp\, dxdt. \tag{2.21}$$

Hence, combining (2.19) and (2.21), we deduce that

$$\int_0^T \int_{\omega_1} (-p + Nu)v\, dxdt = 0,\ \forall v \in L^2(\omega_1^T).$$

Consequently,

$$u = \frac{p}{N} \text{ in } \omega_1^T.$$

Proposition 2. *Let $u \in L^2(\omega_1^T)$ be the solution of (1.4). Then we have the following estimate:*

$$\|u\|_{L^2(\omega_1^T)} \leq C\left(\|h_0\|_{L^2(\mathcal{O}_T)} + \|k\|_{L^2(\omega_2^T)}\right), \tag{2.22}$$

where $C = C(\|a\|_{L^\infty(Q)}, T, N) > 0$.

Proof. Let $z := z(v)$ and $l := l(k)$ be respectively solutions of

$$\begin{cases} -\dfrac{\partial z}{\partial t} - \Delta z + a_0 z &= v\mathbf{1}_{\omega_1} \quad \text{in} \quad Q, \\ z &= 0 \quad \text{on} \quad \Sigma, \\ z(T) &= 0 \quad \text{in} \quad \Omega, \end{cases} \tag{2.23}$$

and

$$\begin{cases} -\dfrac{\partial l}{\partial t} - \Delta l + a_0 l &= h_0\mathbf{1}_{\mathcal{O}} + k\mathbf{1}_{\omega_2} \quad \text{in} \quad Q, \\ l &= 0 \quad \text{on} \quad \Sigma, \\ l(T) &= 0 \quad \text{in} \quad \Omega. \end{cases} \tag{2.24}$$

Then $q = z + l$ and it follows from the data that $z, l \in H^{2,1}(Q)$ and there exist $C = C(\|a\|_{L^\infty(Q)}, T) > 0$ such that

$$\|z\|_{H^{2,1}(Q)} \leq C\|v\|_{L^2(\omega_1^T)} \tag{2.25}$$

and

$$\|l\|_{H^{2,1}(Q)} \leq C\left(\|h_0\|_{L^2(\mathcal{O}_T)} + \|k\|_{L^2(\omega_2^T)}\right). \tag{2.26}$$

Define on $L^2(\omega_1^T) \times L^2(\omega_1^T)$ the bilinear form $a(.,.)$ by

$$a(v,w) = \int_{\Sigma_1} \frac{\partial z(v)}{\partial \nu} \frac{\partial z(w)}{\partial \nu} d\sigma\, dt + N \int_0^T \int_{\omega_1} v\, w dx\, dt \quad \forall v, w \in L^2(\omega_1^T), \tag{2.27}$$

and on $L^2(\omega_1^T)$ the linear form

$$\mathcal{L}(v) = \int_{\Sigma_1} \frac{\partial z(v)}{\partial \nu} \frac{\partial l(k)}{\partial \nu} d\sigma\, dt \quad \forall v \in L^2(\omega_1^T). \tag{2.28}$$

Then, for any $v \in L^2(\omega_1^T)$, the bilinear form $a(.,.)$ is coercive because

$$a(u,u) = \left\| \frac{\partial z(u)}{\partial \nu} \right\|^2_{L^2(\Sigma_1)} + N\|u\|^2_{L^2(\omega_1^T)} \geq N\|u\|^2_{L^2(\omega_1^T)}.$$

In view of (2.19), we have

$$\begin{aligned} 0 &= \int_{\Sigma_1} \frac{\partial z(v)}{\partial \nu} \frac{\partial q}{\partial \nu} d\sigma dt + N \int_0^T \int_{\omega_1} uv dx dt \\ &= \int_{\Sigma_1} \frac{\partial z(v)}{\partial \nu} \frac{\partial z(u)}{\partial \nu} d\sigma dt + \int_{\Sigma_1} \frac{\partial z(v)}{\partial \nu} \frac{\partial l(k)}{\partial \nu} d\sigma dt + N \int_0^T \int_{\omega_1} uv dx dt \\ &= a(v,u) + \int_{\Sigma_1} \frac{\partial z(v)}{\partial \nu} \frac{\partial l(k)}{\partial \nu} d\sigma dt, \forall v \in L^2(\omega_1^T), \end{aligned}$$

where the bilinear form $a(.,.)$ is given by (2.27) and $l = l(k)$ is solution to (2.24). Hence, taking $v = u$ in this latter identity and using the coercivity of $a(.,.)$, we deduce that

$$\begin{aligned} N\|u\|^2_{L^2(\omega_1^T)} &\leq \left\| \frac{\partial z(u)}{\partial \nu} \right\|_{L^2(\Sigma_1)} \left\| \frac{\partial l(k)}{\partial \nu} \right\|_{L^2(\Sigma_1)}, \\ &\leq C\|z(u)\|_{H^{2,1}(Q)} \|l(k)\|_{H^{2,1}(Q)}, \end{aligned}$$

which in view of (2.25) and (2.26) gives

$$\|u\|_{L^2(\omega_1^T)} \leq C \left(\|h_0\|_{L^2(\mathcal{O}_T)} + \|k\|_{L^2(\omega_2^T)} \right),$$

where $C = C(\|a\|_{L^\infty(Q)}, T, N) > 0$. □

3. Carleman Inequality

In order to solve Problem 2, we need an appropriate inequality associated to adjoint of systems (1.7)-(1.9).

Lemma 4. *[10] Let ω_2 be an arbitrary non empty open set of Ω. Then there exists $\psi \in \mathcal{C}^2(\overline{\Omega})$ such that*

$$\psi(x) > 0 \, \forall x \in \Omega, \ \ \psi(x) = 0 \, \forall x \in \Gamma, \tag{3.1}$$

$$|\nabla \psi(x)| > 0 \, \forall x \in \overline{\Omega - \omega_2}. \tag{3.2}$$

Let $\lambda \geq 1$ a real number and m_1 and m_2 be two positive constants such that $m_1 \leq m_2$. For any $(t, x) \in Q$, we set

$$\varphi(t,x) = \frac{e^{\lambda(\psi(x)+m_1)}}{t^2(T-t)^2}, \tag{3.3}$$

$$\eta(t,x) = \frac{e^{\lambda(\psi(x)+m_1)} - e^{\lambda(\|\psi\|_{L^\infty(\Omega)}+m_2)}}{t^2(T-t)^2}. \tag{3.4}$$

Remark 5. *For $x \in \Gamma$, we have $\psi(x) = 0$; therefore, the function φ and η defined respectively by* (3.3) *and* (3.4) *depends only on t and we have*

$$\varphi^{-\frac{1}{4}} \leq C(T).$$

We consider the following system

$$\left\{ \begin{array}{rcll} -\dfrac{\partial z}{\partial t} - \Delta z + a_0 z & = & 0 & \text{in} \quad Q, \\ z & = & h & \text{on} \quad \Sigma, \end{array} \right. \tag{3.5}$$

where $a_0 \in L^\infty(Q)$ and $h \in H^{\frac{1}{4},\frac{1}{2}}(\Sigma)$.
Set

$$W(Q) = \left\{ z | z, \frac{\partial z}{\partial x_i} \in L^2(Q), i = 1, ..., n; \frac{\partial z}{\partial t} \in L^2((0,T); H^{-1}(\Omega)) \right\}.$$

We give now the global Carleman inequality for system (3.5).

Proposition 3. *[11] Let ψ, φ and η be respectively defined by (3.1)-(3.4). Let ω_2 be a non empty open subset of Ω. Then there exists a positive constants $\lambda_0 \geq 1$ and $s_0 \geq 1$ and there exists a constant $C_1 > 0$ independent of $s \geq s_0$ and $\lambda \geq \lambda_0$ such that for any $s \geq s_0$ and $\lambda \geq \lambda_0$, and for all $z \in W(Q)$ solution of (3.5), we have*

$$\begin{aligned}&\int_Q \frac{1}{s\varphi} e^{-2s\eta}|\nabla z|^2 dxdt + s\lambda^2 \int_Q \varphi e^{-2s\eta}|z|^2 dxdt \leq \\ &C_1\left(s^{-\frac{1}{2}}\|\varphi^{-\frac{1}{4}} h e^{-s\eta}\|^2_{H^{\frac{1}{4},\frac{1}{2}}(\Sigma)} + s\lambda^2 \int_0^T \int_{\omega_2} \varphi e^{-2s\eta}|z|^2 dxdt\right).\end{aligned} \tag{3.6}$$

For any suitable function z, we adopt the following notation

$$I(z) = \int_Q \frac{1}{s\varphi} e^{-2s\eta}|\nabla z|^2 dxdt + s\lambda^2 \int_Q e^{-2s\eta}\varphi|z|^2 dxdt. \tag{3.7}$$

For any $\rho^0 \in L^2(\Omega)$, we consider the following systems:

$$\left\{\begin{array}{rcll}\dfrac{\partial \rho}{\partial t} - \Delta\rho + a_0\rho &=& 0 & \text{in } Q,\\ \rho &=& \frac{\partial \Psi}{\partial \nu}\mathbf{1}_{\Gamma_1} & \text{on } \Sigma,\\ \rho(0) &=& \rho^0 & \text{in } \Omega,\end{array}\right. \tag{3.8}$$

$$\left\{\begin{array}{rcll}-\dfrac{\partial \Psi}{\partial t} - \Delta\Psi + a_0\Psi &=& \frac{\rho}{N}\mathbf{1}_{\omega_1} & \text{in } Q,\\ \Psi &=& 0 & \text{on } \Sigma,\\ \Psi(T) &=& 0 & \text{in } \Omega.\end{array}\right. \tag{3.9}$$

Using Proposition 3, we have the following results.

Proposition 4. *Let ω_1 and ω_2 be two non empty open subsets of Ω such that $\omega_1 \subset \omega_2$. Then, there exists a positive constant $C = C(C_1, \|a_0\|_{L^\infty(Q)}, T, N)$ and a weight function θ define by* (3.16) *such that the following observability inequality holds true for any ρ solution of* (3.8)-(3.9)*:*

$$\int_Q e^{-2s\eta}\varphi|\rho|^2 dxdt \leq C \int_0^T \int_{\omega_2} |\rho|^2 dxdt. \tag{3.10}$$

Proof. Applying (3.6) to ρ solution of (3.8)-(3.9), then using the notation (3.7), we have that

$$I(\rho) \leq C_1\left(\left\|s^{-\frac{1}{4}}\varphi^{-\frac{1}{4}}e^{-s\eta}\frac{\partial \Psi}{\partial \nu}\right\|^2_{H^{\frac{1}{4},\frac{1}{2}}(\Sigma_1)} + s\lambda^2 \int_0^T \int_{\omega_2} \varphi e^{-2s\eta}|\rho|^2 dxdt\right). \tag{3.11}$$

In view of the definition of ψ, φ and η given respectively by (3.1), (3.3) and (3.4), we have that there exists $C(T)$ such that

$$\left\| s^{-\frac{1}{4}}\varphi^{-\frac{1}{4}}e^{-s\eta}\frac{\partial\Psi}{\partial\nu}\right\|^2_{H^{\frac{1}{4},\frac{1}{2}}(\Sigma_1)} \leq C(T)\left\|\frac{\partial\Psi}{\partial\nu}\right\|^2_{H^{\frac{1}{4},\frac{1}{2}}(\Sigma_1)}. \tag{3.12}$$

Using on the one hand the fact that

$$\left\|\frac{\partial\Psi}{\partial\nu}\right\|^2_{H^{\frac{1}{4},\frac{1}{2}}(\Sigma_1)} \leq C\|\Psi\|^2_{H^{2,1}(Q)}, \tag{3.13}$$

and, on the other hand that Ψ satisfies (3.9), we deduce that

$$\|\Psi\|^2_{H^{2,1}(Q)} \leq C(\|a_0\|_{L^\infty(Q)}, T, N)\|\rho\|^2_{L^2(\omega_1^T)}. \tag{3.14}$$

Combining (3.12)-(3.14) and using the fact that $\omega_1 \subset \omega_2$, we have that

$$\left\| s^{-\frac{1}{4}}\varphi^{-\frac{1}{4}}e^{-s\eta}\frac{\partial\Psi}{\partial\nu}\right\|^2_{H^{\frac{1}{4},\frac{1}{2}}(\Sigma_1)} \leq C(\|a_0\|_{L^\infty(Q)}, T, N)\|\rho\|^2_{L^2(\omega_2^T)}. \tag{3.15}$$

In view of (3.15) and the fact that $s, \lambda \geq 1$, it follows from (3.11) that

$$\int_Q e^{-2s\eta}\varphi|\rho|^2dxdt \leq C(C_1, \|a_0\|_{L^\infty(Q)}, T, N)\|\rho\|^2_{L^2(\omega_2^T)} + C_1\int_0^T\int_{\omega_2}\varphi e^{-2s\eta}|\rho|^2dxdt.$$

Since $e^{-2s\eta}\varphi \in L^\infty(Q)$, we deduce from this latter inequality (3.10):

$$\int_Q e^{-2s\eta}\varphi|\rho|^2dxdt \leq C(C_1, \|a_0\|_{L^\infty(Q)}, T, N)\|\rho\|^2_{L^2(\omega_2^T)}.$$

Set

$$\frac{1}{\theta^2} = e^{-2s\eta}\varphi. \tag{3.16}$$

Then according to the definition of η and φ, we have that $\frac{1}{\theta^2} \in L^\infty(Q)$ and (3.10) can be rewritten as

$$\int_Q \frac{1}{\theta^2}|\rho|^2dxdt \leq C(C_1, \|a_0\|_{L^\infty(Q)}, T, N)\|\rho\|^2_{L^2(\omega_2^T)}. \tag{3.17}$$

□

4. PROOF OF THEOREM 3

In this section, we are concerned with the Problem 2. More precisely, we are interested in the following null controllability problem: Find a control $k \in L^2(\omega_2^T)$ such that if $q = q(t, x; u(k), k) \in H^{2,1}(Q)$ is solution of (1.7)-(1.9), then (1.6) holds true, i.e.:

$$q(0, .; u(k), k) = 0 \text{ in } \Omega.$$

In other words, we are concerned by the Proof of Theorem 3. To this end, we proceed in three steps.

Step 1. For any $\varepsilon > 0$ and for any $\rho^0 \in L^2(\Omega)$, we consider the following functional:

$$J_\varepsilon(\rho^0) = \frac{1}{2}\int_0^T \int_{\omega_2} |\rho|^2 dxdt + \int_{\mathcal{O}_T} h_0 \rho \, dxdt + \varepsilon \|\rho^0\|_{L^2(\Omega)}, \tag{4.1}$$

where ρ and Ψ are solutions of (3.8) and (3.9). Then we have the following results.

Proposition 5. *Assume that $a_0 \in L^\infty(Q)$ and that $\omega_1 \subset \omega_2$. Let θ be defined by* (3.16) *and $h_0 \in L^2(\mathcal{O}_T)$ be such that $\theta h_0 \in L^2(\mathcal{O}_T)$. Then there exists a unique point $\rho_\varepsilon^0 \in L^2(\Omega)$ such that*

$$J_\varepsilon(\rho_\varepsilon^0) = \inf_{\rho^0 \in L^2(\Omega)} J_\varepsilon(\rho^0). \tag{4.2}$$

Moreover, when $\rho_\varepsilon^0 \neq 0$, we have the following optimality condition

$$\begin{aligned} 0 \quad = \quad & \int_0^T \int_{\omega_2} \rho_\varepsilon \rho dxdt + \int_{\mathcal{O}_T} h_0 \rho dxdt \\ & + \varepsilon \frac{1}{\|\rho_\varepsilon^0\|_{L^2(\Omega)}} \int_\Omega \rho_\varepsilon^0 \rho^0 dx, \; \forall \rho^0 \in L^2(\Omega), \end{aligned} \tag{4.3}$$

where ρ_ε and Ψ_ε are solutions to (3.8) and (3.9) corresponding to $\rho^0 = \rho_\varepsilon^0$. In addition, there exists a positive constant $C = C(C_1, \|a_0\|_{L^\infty(Q)}, T, N)$ independent of ε such that $k_\varepsilon = \rho_\varepsilon \mathbf{1}_{\omega_2}$ satisfies:

$$\|k_\varepsilon\|_{L^2(\omega_2^T)} \leq \sqrt{C}\, \|\theta h_0\|_{L^2(\mathcal{O}_T)}. \tag{4.4}$$

Proof. It is clear that J_ε is continuous and strictly convex on $L^2(\Omega)$. It remains to prove that J_ε is coercive to obtain that there exists a unique point $\rho^0_\varepsilon \in L^2(\Omega)$ where the functional J_ε reaches its minimum. So, using Cauchy Schwarz inequality and the definition of θ given by (3.16), we have that

$$\int_Q (-h_0)\rho\, dxdt \leq \|\theta h_0\|_{L^2(\mathcal{O}_T)} \left\|\frac{1}{\theta}\rho\right\|_{L^2(Q)},$$

which in view of (3.10) gives

$$\int_Q (-h_0)\rho\, dxdt \leq \sqrt{C}\, \|\theta h_0\|_{L^2(\mathcal{O}_T)} \|\rho\|_{L^2(\omega_2^T)}$$

where $C = C(C_1, \|a_0\|_{L^\infty(Q)}, T, N) > 0$. Using Young inequality, we obtain that

$$\int_Q (-h_0)\rho\, dxdt \leq C\|\theta h_0\|^2_{L^2(\mathcal{O}_T)} + \frac{1}{4}\|\rho\|^2_{L^2(\omega_2^T)},$$

which combining with definition of J_ε given by (4.1) yields

$$J_\varepsilon(\rho^0) \geq \frac{1}{4}\int_0^T \int_{\omega_2} |\rho|^2 dxdt + \varepsilon\|\rho^0\|_{L^2(\Omega)} - C\|\theta h_0\|^2_{L^2(\mathcal{O}_T)}$$

where $C = C(C_1, \|a_0\|_{L^\infty(Q)}, T, N) > 0$. Hence we deduce that J_ε is coercive on $L^2(\Omega)$. Consequently, there exists a unique point $\rho^0_\varepsilon \in L^2(\Omega)$ where the functional J_ε reaches its minimum.

Now, assume that $\rho^0_\varepsilon \neq 0$. In order to prove (4.3), we write the Euler-Lagrange condition which characterize the minimizer $\rho^0_\varepsilon \in L^2(\Omega)$:

$$\lim_{\lambda \to 0} \frac{J_\varepsilon(\rho^0_\varepsilon + \lambda\rho^0) - J_\varepsilon(\rho^0_\varepsilon)}{\lambda} = 0,\ \forall \rho^0 \in L^2(\Omega). \tag{4.5}$$

After some calculations (4.5) yields (4.3).

Let ρ^0_ε be the solution of (4.2) and ρ_ε be the solution of (3.8)-(3.9) associated to ρ^0_ε. Let also q_ε, p_ε be the solution associated to $k = k_\varepsilon = \rho_\varepsilon$ of systems (1.7) and (1.8) respectively:

$$\left\{\begin{array}{rcll} -\dfrac{\partial q_\varepsilon}{\partial t} - \Delta q_\varepsilon + a_0 q_\varepsilon &=& h_0\mathbf{1}_{\mathcal{O}} + u_\varepsilon\mathbf{1}_{\omega_1} + k_\varepsilon\mathbf{1}_{\omega_2} & \text{in } Q,\\ q_\varepsilon &=& 0 & \text{on } \Sigma,\\ q_\varepsilon(T) &=& 0 & \text{in } \Omega, \end{array}\right. \tag{4.6}$$

and

$$\left\{\begin{array}{rcll}\dfrac{\partial p_\varepsilon}{\partial t}-\Delta p_\varepsilon+a_0p_\varepsilon & = & 0 & \text{in} \quad Q,\\ p_\varepsilon & = & \frac{\partial q_\varepsilon}{\partial\nu}\mathbf{1}_{\Gamma_1} & \text{on} \quad \Sigma,\\ p_\varepsilon(0) & = & 0 & \text{in} \quad \Omega,\end{array}\right. \tag{4.7}$$

with

$$u_\varepsilon=\frac{p_\varepsilon}{N}\ \text{in}\ \omega_1^T, \tag{4.8}$$

and

$$k_\varepsilon=\rho_\varepsilon\mathbf{1}_{\omega_2}\ \text{in}\ Q. \tag{4.9}$$

Multiplying the first line of (4.6) by ρ and the first line of (4.7) by Ψ, where ρ and Ψ are respectively solutions of (3.8) and (3.9) and integrating by parts over Q, we have successively

$$\int_\Omega q_\varepsilon(0)\rho^0\,dx-\int_{\Sigma_1}\frac{\partial q_\varepsilon}{\partial\nu}\frac{\partial\Psi}{\partial\nu}\,d\sigma dt=\int_{\mathcal{O}_T}h_0\rho\,dxdt+\frac{1}{N}\int_{\omega_1^T}p_\varepsilon\rho\,dxdt+\int_{\omega_2^T}\rho_\varepsilon\rho\,dxdt \tag{4.10}$$

and

$$\int_{\Sigma_1}\frac{\partial q_\varepsilon}{\partial\nu}\frac{\partial\Psi}{\partial\nu}\,d\sigma dt+\frac{1}{N}\int_{\omega_1^T}p_\varepsilon\rho\,dxdt=0. \tag{4.11}$$

Combining (4.10) and (4.11) with (4.3), we obtain

$$0=\int_\Omega\left(q_\varepsilon(0,x)+\varepsilon\frac{1}{\|\rho_\varepsilon^0\|_{L^2(\Omega)}}\rho_\varepsilon^0(x)\right)\rho^0(x)dx,\ \forall\rho^0\in L^2(\Omega).$$

Hence,

$$q_\varepsilon(0,x)=-\varepsilon\frac{1}{\|\rho_\varepsilon^0\|_{L^2(\Omega)}}\rho_\varepsilon^0(x).$$

Consequently,

$$\|q_\varepsilon(0,x)\|_{L^2(\Omega)}=\varepsilon. \tag{4.12}$$

Now, if we take $\rho^0=\rho_\varepsilon^0$ in (4.3), we obtain that

$$\begin{aligned}\|k_\varepsilon\|^2_{L^2(\omega_2^T)} &= \int_0^T\int_\omega|\rho_\varepsilon|^2dxdt=-\int_{\mathcal{O}_T}h_0\rho_\varepsilon\,dxdt-\varepsilon\|\rho_\varepsilon^0\|_{L^2(\Omega)}\\ &\le \|\theta h_0\|_{L^2(\mathcal{O}_T)}\left\|\frac{1}{\theta}\rho_\varepsilon\right\|_{L^2(Q)}.\end{aligned}$$

It then follows from (3.10) that

$$\|k_\varepsilon\|_{L^2(\omega_2^T)}\le\sqrt{C}\,\|\theta h_0\|_{L^2(\mathcal{O}_T)},$$

where $C=C(C_1,\|a_0\|_{L^\infty(Q)},T,N)>0$. □

Step 2. We give some a priori estimates on u_ε, q_ε and p_ε.

In view of (2.22) and (4.4), we have that

$$\|u_\varepsilon\|_{L^2(\omega_2^T)} \leq C\left(\|h_0\|_{L^2(\mathcal{O}_T)} + \|\theta h_0\|_{L^2(\mathcal{O}_T)}\right), \tag{4.13}$$

where $C = C(C_1, \|a_0\|_{L^\infty(Q)}, T, N) > 0$. Since $q_\varepsilon, p_\varepsilon$ satisfy (4.6)-(4.7), using (4.4) and (4.13), we prove that

$$\|q_\varepsilon\|_{H^{2,1}(Q)} \leq C\left(\|h_0\|_{L^2(\mathcal{O}_T)} + \|\theta h_0\|_{L^2(\mathcal{O}_T)}\right), \tag{4.14}$$

$$\left\|\frac{\partial q_\varepsilon}{\partial \nu}\right\|_{L^2(\Sigma)} \leq C\left(\|h_0\|_{L^2(\mathcal{O}_T)} + \|\theta h_0\|_{L^2(\mathcal{O}_T)}\right), \tag{4.15}$$

and

$$\|p_\varepsilon\|_{L^2(Q)} \leq C\left(\|h_0\|_{L^2(\mathcal{O}_T)} + \|\theta h_0\|_{L^2(\mathcal{O}_T)}\right), \tag{4.16}$$

where $C = C(C_1, \|a_0\|_{L^\infty(Q)}, T, N) > 0$.

Step 3. We study the convergence when $\varepsilon \to 0$ to the sequences k_ε, u_ε, q_ε, $\frac{\partial q_\varepsilon}{\partial \nu}$ and p_ε.

In view of (4.4), (4.12), (4.13)-(4.16), we can extract subsequences still denoted by k_ε, u_ε, q_ε, $\frac{\partial q_\varepsilon}{\partial \nu}$ and p_ε such that when $\varepsilon \to 0$, we have

$$k_\varepsilon \rightharpoonup \hat{k} \text{ weakly in } L^2(\omega_2^T), \tag{4.17a}$$

$$u_\varepsilon \rightharpoonup u \text{ weakly in } L^2(\omega_1^T), \tag{4.17b}$$

$$q_\varepsilon \rightharpoonup q \text{ weakly in } H^{2,1}(Q), \tag{4.17c}$$

$$\frac{\partial q_\varepsilon}{\partial \nu} \rightharpoonup \beta \text{ weakly in } L^2(\Sigma), \tag{4.17d}$$

$$p_\varepsilon \rightharpoonup p \text{ weakly in } L^2(Q), \tag{4.17e}$$

$$q_\varepsilon(0) \to 0 \text{ strongly in } L^2(\Omega). \tag{4.17f}$$

From (4.8), (4.17b) and (4.17e), we obtain

$$u = \frac{p}{N} \text{ in } \omega_1^T. \tag{4.18}$$

Moreover using the weak lower semi-continuity of the norm, we deduce from (4.17b) and (4.13) that

$$\|u\|_{L^2(\omega_1^T)} \leq C(C_1, \|a_0\|_{L^\infty(Q)}, T, N)\left(\|h_0\|_{L^2(\mathcal{O}_T)} + \|\theta h_0\|_{L^2(\mathcal{O}_T)}\right). \tag{4.19}$$

Now, proceeding as for the convergence of q_n in page 132-134, we prove using (4.17a)-(4.17d) and (4.17f) that q is solution of

$$\begin{cases} -\dfrac{\partial q}{\partial t} - \Delta q + a_0 q &= h_0 \mathbf{1}_{\mathcal{O}} + u\mathbf{1}_{\omega_1} + \hat{k}\mathbf{1}_{\omega_2} \quad \text{in} \quad Q, \\ q &= 0 \quad \text{on} \quad \Sigma, \\ q(T) &= 0 \quad \text{in} \quad \Omega, \end{cases} \tag{4.20}$$

$$\frac{\partial q_\varepsilon}{\partial \nu} \rightharpoonup \beta = \frac{\partial q}{\partial \nu} \text{ weakly in } L^2(\Sigma) \tag{4.21}$$

and

$$q(0) = 0 \text{ in } \Omega. \tag{4.22}$$

Now, if we multiply the first equation in (4.7) by $\xi \in \mathbb{D}(Q)$ and integrate by parts over Q, then take the limit when $\varepsilon \to 0$ while using (4.17), we deduce that

$$\int_Q p\left(-\frac{\partial \xi}{\partial t} - \Delta\xi + a_0\xi\right) dxdt = 0,$$

which after an integration by parts over Q gives

$$\int_Q \left(\frac{\partial p}{\partial t} - \Delta p + a_0 p\right) \xi \, dxdt = 0, \ \forall \xi \in \mathbb{D}(Q).$$

Hence, we deduce that

$$-\frac{\partial p}{\partial t} - \Delta p + a_0 p = 0 \text{ in } Q. \tag{4.23}$$

Since $p \in L^2(Q)$ and $\Delta p \in H^{-1}((0,T); L^2(\Omega))$, we deduce that $p|_\Sigma$ and $\dfrac{\partial p}{\partial \nu}|_\Sigma$ exist and belong to $H^{-1}((0,T); H^{-1/2}(\Gamma))$ and $H^{-1}((0,T); H^{-3/2}(\Gamma))$ respectively. On the other hand observing that $p \in L^2(Q)$ and $\dfrac{\partial p}{\partial t} \in L^2((0,T); H^{-2}(\Omega))$, we deduce that $p(0)$ and $p(T)$ exist in $H^{-1}(\Omega)$.

So, if we multiply the first equation in (4.7) by $\xi \in \mathcal{C}^\infty(\overline{Q})$ such that $\xi|_\Sigma = 0$, $\xi(T) = 0$ in Ω, then integrate by parts over Q, we obtain

$$0 = \int_{\Sigma_1} \frac{\partial q_\varepsilon}{\partial \nu}\frac{\partial \xi}{\partial \nu} \, d\sigma dt + \int_Q p_\varepsilon \left(-\frac{\partial \xi}{\partial t} - \Delta\xi + a_0\xi\right) dxdt,$$
$$\forall \xi \in \mathcal{C}^\infty(\overline{Q}) \text{ such that } \xi|_\Sigma = 0, \ \xi(T) = 0 \text{ in } \Omega.$$

Passing this latter identity to the limit when ε tends toward zero, while using (4.17e) and (4.21), we obtain that

$$0 = \int_{\Sigma_1} \frac{\partial q}{\partial \nu}\frac{\partial \xi}{\partial \nu}\, d\sigma dt + \int_Q p\left(-\frac{\partial \xi}{\partial t} - \Delta\xi + a_0\xi\right) dxdt,$$
$$\forall \xi \in \mathcal{C}^\infty(\overline{Q}) \text{ such that } \xi|_\Sigma = 0,\ \xi(T) = 0 \text{ in } \Omega,$$

which after an integration by parts yields

$$\begin{aligned}
&0 = \langle p(0), \xi(0)\rangle_{H^{-1}(\Omega), H_0^1(\Omega)} + \int_{\Sigma_1} \frac{\partial q}{\partial \nu}\frac{\partial \xi}{\partial \nu}\, d\sigma dt - \\
&\left\langle p, \frac{\partial \xi}{\partial \nu}\right\rangle_{H^{-3/2}((0,T), H^{-1/2}(\Gamma)),\, H^{3/2}((0,T), H^{1/2}(\Gamma))} \\
&+ \int_Q \left(\frac{\partial p}{\partial t} - \Delta p + a_0 p\right) \xi\, dxdt, \\
&\forall \xi \in \mathcal{C}^\infty(\overline{Q}) \text{ such that } \xi|_\Sigma = 0,\ \xi(T) = 0 \text{ in } \Omega.
\end{aligned}$$

Using (4.23) in this latter identity, we get

$$\begin{aligned}
&0 = \langle p(0), \xi(0)\rangle_{H^{-1}(\Omega), H_0^1(\Omega)} + \int_{\Sigma_1} \frac{\partial q}{\partial \nu}\frac{\partial \xi}{\partial \nu}\, d\sigma dt - \\
&\left\langle p, \frac{\partial \xi}{\partial \nu}\right\rangle_{H^{-3/2}((0,T), H^{-1/2}(\Gamma)),\, H^{3/2}((0,T), H^{1/2}(\Gamma))} \\
&\forall \xi \in \mathcal{C}^\infty(\overline{Q}) \text{ such that } \xi|_\Sigma = 0,\ \xi(T) = 0 \text{ in } \Omega.
\end{aligned} \tag{4.24}$$

Take in (4.24), ξ such that $\frac{\partial \xi}{\partial \nu} = 0$ on Σ, we deduce that

$$p(0) = 0 \text{ in } \Omega, \tag{4.25}$$

and finally,

$$p = \frac{\partial q}{\partial \nu} \text{ on } \Sigma_1. \tag{4.26}$$

It then follows from (4.23), (4.26) and (4.25) that p satisfies

$$\left\{\begin{array}{rcll}
\dfrac{\partial p}{\partial t} - \Delta p + a_0 p &=& 0 & \text{in } Q, \\
p &=& \frac{\partial q}{\partial \nu}\mathbf{1}_{\Gamma_1} & \text{on } \Sigma, \\
p(0) &=& 0 & \text{in } \Omega.
\end{array}\right. \tag{4.27}$$

Finally, using the weak-lower semi-continuity of the norm and (4.17a), w deduce from (4.4) the estimate (1.14).

Observing that $k_\varepsilon = \rho_\varepsilon \mathbf{1}_{\omega_2}$, where ρ_ε satisfies:

$$\left\{\begin{array}{rcll} \dfrac{\partial \rho_\varepsilon}{\partial t} - \Delta \rho_\varepsilon + a_0 \rho_\varepsilon &=& 0 & \text{in } Q, \\ \rho_\varepsilon &=& \frac{\partial \Psi_\varepsilon}{\partial \nu} \mathbf{1}_{\Gamma_1} & \text{on } \Sigma, \\ \rho_\varepsilon(0) &=& \rho_\varepsilon^0 & \text{in } \Omega, \end{array}\right. \tag{4.28}$$

with Ψ_ε solution of

$$\left\{\begin{array}{rcll} -\dfrac{\partial \Psi_\varepsilon}{\partial t} - \Delta \Psi_\varepsilon + a_0 \Psi_\varepsilon &=& -\frac{\rho_\varepsilon}{N} \mathbf{1}_{\omega_1} & \text{in } Q, \\ \Psi_\varepsilon &=& 0 & \text{on } \Sigma, \\ \Psi_\varepsilon(T) &=& 0 & \text{in } \Omega, \end{array}\right. \tag{4.29}$$

we know that

$$\begin{array}{rcl} \|\Psi_\varepsilon\|_{H^{2,1}(Q)} &\leq& C(\|a_0\|_{L^\infty(Q)}, T, N) \|\rho_\varepsilon\|_{L^2(\omega_2)} \\ &\leq& C(C_1, \|a_0\|_{L^\infty(Q)}, T, N) \|\theta h_0\|_{L^2(\mathcal{O}_T)}, \\ \left\|\dfrac{\partial \Psi_\varepsilon}{\partial \nu}\right\|_{L^2(\Sigma)} &\leq& C(C_1, \|a_0\|_{L^\infty(Q)}, T, N) \|\theta h_0\|_{L^2(\mathcal{O}_T)}, \end{array}$$

because $\omega_1 \subset \omega_2$, $k_\varepsilon = \rho_\varepsilon \mathbf{1}_{\omega_2}$ and (4.4) holds true.

Consequently, there exist $\hat{\Psi} \in H^{2,1}(Q)$ and $\beta \in L^2(\Sigma)$ such that

$$\Psi_\varepsilon \rightharpoonup \hat{\Psi} \text{ weakly in } H^{2,1}(Q), \tag{4.30a}$$

$$\frac{\partial \Psi_\varepsilon}{\partial \nu} \rightharpoonup \beta \text{ weakly in } \quad L^2(\Sigma). \tag{4.30b}$$

On the other hand, from (3.10), (4.4) and (4.9), there exists $C = C(C_1, \|a_0\|_{L^\infty(Q)}, T, N) > 0$ such that

$$\left\|\frac{1}{\theta}\rho_\varepsilon\right\|^2_{L^2(Q)} \leq C\, \|\theta h_0\|^2_{L^2(\mathcal{O}_T)}.$$

Hence, if we set $L^2(\frac{1}{\theta}, X) = \left\{ z \in L^2(X), \displaystyle\int_X \frac{1}{\theta^2} |z|^2 dX < \infty \right\}$, we deduce from this latter inequality that ρ_ε is bounded in $L^2(\frac{1}{\theta}, Q)$. Consequently, there exists $\hat{\rho} \in L^2\left(\frac{1}{\theta}, Q\right)$ and a subsequence of (ρ_ε) still denoted (ρ_ε) such that

$$\rho_\varepsilon \rightharpoonup \hat{\rho} \text{ weakly in } L^2\left(\frac{1}{\theta}, Q\right). \tag{4.31}$$

If we refer of the definition of ψ, φ and η given by (3.1)-(3.4) and the definition of θ given by (3.16), we can see that for all $\tau > 0$,

$$\rho_\varepsilon \rightharpoonup \hat{\rho} \text{ weakly in } L^2(]\tau, T-\tau[\times\Omega).$$

This implies that

$$\rho_\varepsilon \rightharpoonup \hat{\rho} \text{ weakly in } \mathbb{D}'(Q). \tag{4.32}$$

where $\mathbb{D}'(Q)$ is the dual of $\mathbb{D}(Q)$. Therefore, it follows from (4.4) and (4.9) that

$$\rho_\varepsilon \rightharpoonup \hat{\rho} \text{ weakly in } L^2(\omega_2^T) \tag{4.33}$$

and we have

$$\hat{k} = \hat{\rho}\mathbf{1}_{\omega_2}. \tag{4.34}$$

Now proceeding as for the convergence of q_n in page 132-134, while passing to the limit in (4.29), we prove using (4.30) and (4.33) that $\hat{\Psi}$ is solution of (1.13) and that

$$\frac{\partial \Psi_\varepsilon}{\partial \nu} \rightharpoonup \frac{\partial \hat{\Psi}}{\partial \nu} \text{ weakly in } L^2(\Sigma). \tag{4.35}$$

Using (4.32) and (4.35) while passing to the limit in (4.28), we obtain that $\hat{\rho}$ is solution of (1.12).

In short, In view of (4.20), (4.22) and (4.27), we have that $(\hat{k}, q, p)$ is solution of the null controllability problems (1.7)-(1.9) and (1.6).

Conclusion

In this paper we proved that (1.1) is controllable in the sense of Stackelberg. The nonlinear case will be investigated in our future work.

Acknowledgments

The first author was supported by the German Academic Exchange Service (D.A.A.D) under the Scholarship Program PhD AIMS-Cameroon.
The second author was supported by the Alexander von Humboldt Foundation under the program financed by the BMBF entitled “German research Chairs”.

REFERENCES

[1] Lions J. L., Magenes E., *Problèmes aux Limites non Homogènes et Applications*, [*Non-homogeneous boundary problems and applications*, Dunod, Paris, (1968)].

[2] Von Stackelberg H., *Markform und Gleichgewicht*, [*Mark shape and balance*, Springer, Berlin, Germany (4) (1934)].

[3] Lions J. L., Some remarks on Stackelberg's optimization, *Mathematical Model and Methods in Applied Sciences* 4 (4) (1994), pp 477-487.

[4] Nakoulima O., Optimal control for distributed systems subject to null controllability. Application to discriminating sentinels, *ESAIM: COCV* 13 (4) (2007), pp 623-638.

[5] Mercan M., Optimal control for distributed linear systems subjected to null-controllability, *Applicable Analysis* 92 (9) (2013), pp 1928-1943.

[6] Mercan M., Optimal Control for Distributed Linear Systems Subjected to Null Controllability with Constraints on the State, *Springer Proceedings in Mathematics and Statistics* 37 (2013), pp 213-232.

[7] Mercan M., Nakoulima O., Control of Stackelberg for a two stroke problem, Dynamics of continuous, *Discrete and Impulsive systems: Applications and Algorithms* 22 (2015), pp 441-463.

[8] Araruna F. D., Fernández-Cara E., Santos M. C., Stackelberg-Nash exact controllabillity for linear and semilinear parabolic equations, *ESAIM: COCV* 21 (3) (2015), pp 835-856.

[9] Hernández-Santamaría V., De Teresa L., Robust Stackelberg controllability for linear and semilinear heat equations, *Evolution Equation and Control Theory* 7 (2) (2018), pp 247-273.

[10] Fursikov A. V., Imanuvilov Yu O., *Controllability of Evolution Equations, Lecture Notes*, Research Institute of Mathematics, Seoul National University, Seoul, Korea 34 (1994).

[11] Imanuvilov Yu O., Puel J. P., Yamamoto M., Carleman estimates for a parabolic equations with non homogeneous boundary conditions, *Chin. Annal. Math 30* (4) (2009), pp 333-378.

In: Evolutionary Processes and Applications
Editor: G. M. N'Guérékata
ISBN: 978-1-53616-350-6

Chapter 9

S-ASYMPTOTICALLY ω-PERIODIC MILD SOLUTIONS FOR SECOND ORDER EVOLUTION EQUATIONS

***Mouffak Benchohra*[1,*], *Gaston M. N'Guérékata*[2,†] *and Noreddine Rezoug*[3,‡]**
[1]Laboratory of Mathematics, Djillali Liabes,
University of Sidi Bel-Abbes, Sidi Bel-Abbes, Algeria
[2]Department of Mathematics,
Morgan State University, Baltimore, MD, US
[3]Département de Mathématiques, Centre Universitaire de Relizane,
Relizane, Algeria

Abstract

This chapter is devoted to study the existence of S-asymptotically ω-periodic mild solution to some classes of second order semilinear evolution equation in Banach space, via the techniques of measure of noncompactness. The investigation is based on a new fixed point result which is a generalization of the well known Darbo's fixed point theorem. Finally, an example is presented to illustrate the main findings.

*E-mail address: benchohra@yahoo.com.
†E-mail address: Gaston.N'Guerekata@morgan.edu.
‡E-mail address: noreddinerezoug@yahoo.fr.

Keywords: second order differential equations, mild solution, ω-periodic, S-asymptotically ω-periodic, evolution system, Kuratowski measures of noncompactness, fixed point

1. INTRODUCTION

In this chapter, we investigate the existence of S-asymptotically ω-periodic mild solution for second differential equations. More precisely, we will consider the following problem

$$y''(t) - A(t)y(t) = f(t, y(t)),\ t \in J := [0, \infty), \tag{1.1}$$

$$y(0) = y_0,\ y'(0) = y_1, \tag{1.2}$$

where $\{A(t)\}_{0 \le t < +\infty}$ is a family of linear closed operators from E into E that generate an evolution system of linear bounded operators $\{U(t,s)\}_{(t,s)\in J\times J}$ for $0 \le s \le t < +\infty$, $f : J \times E \to E$ is a Carathéodory function, and $(E, |\cdot|)$ is a real Banach space.

Evolution equations arise in many areas of applied mathematics [2, 36]. This type of equations have received much attention in recent years [1]. There are many results concerning the second-order differential equations, see for example [7, 10, 20, 26, 34]. In recent years there has been an increasing interest in studying the abstract non-autonomous second order initial value problem

$$y''(t) - A(t)y(t) = f(t, y(t)),\ t \in [0, T], \tag{1.3}$$

$$y(0) = y_0,\ y'(0) = y_1. \tag{1.4}$$

The reader is referred to [9, 19, 23, 35] and the references therein. In the above mentioned works, the existence of solutions to the problem (1.3)-(1.4) is related to the existence of an evolution operator $U(t; s)$ for the homogeneous equation

$$y''(t) = A(t)y(t), \quad \text{for } t \ge 0.$$

For this purpose there are many techniques to show the existence of $U(t, s)$ which has been developed by Kozak [25].

Actually, while the almost periodic, almost automorphic, and weighted pseudo almost periodic solutions to various evolution equations are investigated by many authors ([16, 30, 33, 37]), the notion of S-asymptotically ω-periodic functions have many applications in several problems like functional differential equations, integro-differential equations, fractional differential equations and fractional integro-differential equations. The concept of S-asymptotically ω-periodic function was first introduced in the literature by Henriquez *et al.* in [21, 22]. In the literature, there has been a significant attention devoted this concept in the deterministic case; we refer the reader to [11, 13, 14, 15, 17, 18, 28, 29] and the references therein.

In this work we use the technique of measures of noncompactness. It is well known that this method provides an excellent tool for obtaining existence of solutions of nonlinear differential equation. This technique works fruitfully for both integral and differential equations. More details are found in Aissani and Benchohra [3], Akhmerov *et al.* [4], Alv́ares [5], Banaś and Goebel [8], Guo *et al.* [27], Olszowy and Wędrychowicz [31], Olszowy [32], and the references therein.

Inspired by the above works, in this chapter, we establish sufficient conditions for the existence of solutions of the system (1.1)-(1.2) by means the Kuratowski measure of noncompactness and the fixed point approach.

This work is organized of as follows. In Section 2, we recall some definitions and facts about evolution systems. In Section 3, we give the existence of mild solutions to the problem (1.1)-(1.2). Section 4 is devoted to an example illustrating the abstract theory.

2. PRELIMINARIES

In this section, we mention notations, definitions, lemmas and preliminary facts needed to establish our main results.

Throughout this chapter, we denote by E a Banach space with the norm $|\cdot|$. Let $BC(J, E)$ be the Banach space of all bounded and continuous functions y mapping J into E with the usual supremum norm

$$\|y\|_\infty = \sup_{t\in J} |y(t)|.$$

In what follows, let $\{A(t),\ t \geq 0\}$ be a family of closed linear operators on the Banach space E with domain $D(A(t))$ which is dense in E and independent of t.

In this work, the existence of solution the problem (1.1)-(1.2) is related to the existence of an evolution operator $U(t, s)$ for the following homogeneous problem

$$y''(t) = A(t)y(t) \qquad t \in J. \tag{2.1}$$

This concept of evolution operator has been developed by Kozak [25].

Definition 1. *A family U of linear bounded operators $U(t,s) : E \to E$, $(t,s) \in \Delta := \{(t,s) \in J \times J : s \leq t\}$, is called an evolution operator of the equation (2.1) if de following conditions hold:*

(e_1) *For any $x \in E$ the map $(t,s) \longmapsto U(t,s)x$ is continuously differentiable and*

(a) *for any $t \in J$, $U(t,t) = 0$,*

(b) *for all $(t,s) \in \Delta$ and for any $x \in E$, $\frac{\partial}{\partial t}U(t,s)x\big|_{t=s} = x$ and $\frac{\partial}{\partial s}U(t,s)x\big|_{t=s} = -x$.*

(e_2) *For all $(t,s) \in \Delta$, if $x \in D(A(t))$, then $\frac{\partial}{\partial s}U(t,s)x \in D(A(t))$, the map $(t,s) \longmapsto U(t,s)x$ is of class C^2 and*

(a) $\frac{\partial^2}{\partial t^2}U(t,s)x = A(t)U(t,s)x,$

(b) $\frac{\partial^2}{\partial s^2}U(t,s)x = U(t,s)A(s)x,$

(c) $\frac{\partial^2}{\partial s \partial t}U(t,s)x\big|_{t=s} = 0.$

(e_3) *For all $(t,s) \in \Delta$, then $\frac{\partial}{\partial s}U(t,s)x \in D(A(t))$, there exist $\frac{\partial^3}{\partial t^2 \partial s}U(t,s)x$, $\frac{\partial^3}{\partial s^2 \partial t}U(t,s)x$ and*

(a) $\frac{\partial^3}{\partial t^2 \partial s}U(t,s)x = A(t)\frac{\partial}{\partial s}(t)U(t,s)x.$

Moreover, the map $(t,s) \longmapsto A(t)\frac{\partial}{\partial s}(t)U(t,s)x$ is continuous,

(b) $\frac{\partial^3}{\partial s^2 \partial t}U(t,s)x = \frac{\partial}{\partial t}U(t,s)A(s)x.$

Throughout this work, we will use the following definition of the concept of Kuratowski measure of noncompactness [8].

Definition 2. *The Kuratowski measure of noncompactness α is defined by*

$$\alpha(D) = \inf\{r > 0 : D \text{ has a finite cover by sets of diameter} \leq r\},$$

for a bounded set D in any Banach space X.

Let us recall the basic properties of Kuratowski measure of noncompactness.

Lemma 3. *[8] Let X be a Banach space and $C, D \subset X$ be bounded subsets, then the following properties hold:*

(i$_1$) $\alpha(D) = 0$ *if and only if D is relatively compact,*

(i$_2$) $\alpha(\overline{D}) = \alpha(D)$; $\overline{D}$ *the closure of D,*

(i$_3$) $\alpha(C) \leq \alpha(D)$ *when $C \subset D$,*

(i$_4$) $\alpha(C + D) \leq \alpha(C) + \alpha(D)$ *where* $C + D = \{x \mid x = y + z; y \in C; z \in D\}$,

(i$_5$) $\alpha(aD) = |a|\alpha(D)$ *for any $a \in \mathbb{R}$,*

(i$_6$) $\alpha(ConvD) = \alpha(D)$, *where $ConvD$ is the convex hull of D,*

(i$_7$) $\mu(C \cup D) = \max(\alpha(C), \alpha(D))$,

(i$_8$) $\alpha(C \cup \{x\}) = \alpha(C)$ *for any $x \in X$.*

Denote by $\omega^T(y, \varepsilon)$ the modulus of continuity of y on the interval $[0, T]$ i.e.

$$\omega^T(y, \varepsilon) = \sup\left\{|y(t) - y(s)| ; t, s \in [0, T], |t - s| \leq \varepsilon\right\}.$$

Moreover, let us put

$$\omega^T(D, \varepsilon) = \sup\left\{\omega^T(y, \varepsilon); y \in D\right\},$$

$$\omega_0^T(D) = \lim_{\varepsilon \to 0} \omega^T(D, \varepsilon).$$

Lemma 4. *[12] Let X be a Banach space, $D \subset X$ be bounded. Then there exists a countable set $D_0 \subset D$, such that*

$$\alpha(D) \leq 2\alpha(D_0).$$

Lemma 5. *[24] Let X be a Banach space, $D = \{y_n\}_{n=0}^{+\infty} \subset C([0,T],X)$ be a bounded and countable set. Then $\alpha(D(t))$ is Lebesgue integrable on $[0,T]$, and*

$$\alpha\left(\left\{\int_0^t y_n(s))ds\right\}_{n=0}^{\infty}\right) \leq 2\int_0^t \alpha(D(s))ds, \qquad t \in [0,T].$$

Definition 6. *[22] A function $f \in BC(\mathbb{R}^+, X)$ is called S-asymptotically ω-periodic if there exists $\omega > 0$ such that $\lim_{t\to+\infty}(f(t+\omega) - f(t)) = 0$. In this case we say that ω is an asymptotic period of f and that f is S-asymptotically ω-periodic.*

We will denote by $SAP_\omega(X)$, the set of all S-asymptotically ω-periodic functions from $\mathbb{R}^+$ to X.

The following result is due to Henriquez-Pierri-Táboas; Proposition 3.5 in [21].

Theorem 7. *Endowed with the norm $\|y\|_\infty$, $SAP_\omega(X)$ is a Banach space.*

Sketch of the proof. We consider the translation operator $\tau_\omega : BC(\mathbb{R}^+, X) \to BC(\mathbb{R}^+, X)$ defined by $\tau_\omega f(t) = f(t+\omega)$. Clearly the function τ_ω is linear and continuous. We note that $SAP_\omega(X) = (\tau_\omega - I)^{-1}(C_0(\mathbb{R}^+, X))$. And then, since $(\tau_\omega - I)$ is linear continuous and since $C_0(\mathbb{R}^+, X)$ is a closed vector subspace of $BC(\mathbb{R}^+, X)$, $SAP_\omega(X)$ is a closed vector subspace of the Banach space $BC(\mathbb{R}^+, X)$.

Theorem 8. *[6] Let Ω be a nonempty, bounded, closed and convex subset of a Banach space X, and let $\Gamma : \Omega \to \Omega$ be a continuous operator satisfying the inequality*

$$\alpha(\Gamma(D) \leq \Psi(\alpha(D))$$

for any nonempty subset D of Ω, where $\Psi : \mathbb{R}^+ \to \mathbb{R}^+$ is a nondecreasing function such that

$$\lim_{n\to+\infty} \Psi^n(t) = 0, \textit{ for each } t \geq 0.$$

Then Γ has at least one fixed point in the set Ω.

3. MAIN RESULTS

Definition 9. *A function $y \in BC(J, E)$ is called a mild solution to the problem (1.1)-(1.2) if y satisfies the integral equation*

$$y(t) = -\frac{\partial}{\partial s}U(t,0)y_0 + U(t,0)y_1 + \int_0^t U(t,s)f(s,y(s))ds. \tag{3.1}$$

For the proof of our main theorem, we need the following hypotheses:

(H_1) There exist constants $M \geq 1$ and $\delta > 0$, such that

$$\|U(t,s)\|_{B(E)} \leq Me^{-\delta(t-s)} \quad \text{for any } (t,s) \in \Delta$$

and

$$U(t+\omega, s+\omega) = U(t,s)(\omega - \text{periodicity}).$$

(H_2) There exist constants $\tilde{M} \geq 0$ and $\delta > 0$, such that:

$$\|\frac{\partial}{\partial s}U(t,s)\|_{B(E)} \leq \tilde{M}e^{-\delta(t-s)}, (t,s) \in \Delta.$$

(H_3) The function $f : J \times E \to E$ is Carathéodory and satisfies the following:

(a) There exists a constant $\omega > 0$, such that

$$\lim_{t\to+\infty} |f(t+\omega, u) - f(t,u)| = 0 \text{ for a.e } t \in J \text{ and each } u \in E.$$

(b) There exist $p \in L^q(J, \mathbb{R}^+)$, $q \in (1, \infty)$ and a continuous nondecreasing function $\psi : [0, \infty) \to (0, \infty)$ such that:

$$|f(t,u)| \leq p(t)\psi(|u|) \text{ for a.e } t \in J \text{ and each } u \in E.$$

(H_4) There exist a locally integrable function $\eta : J \to \mathbb{R}^+$ and a continuous nondecreasing function $\varphi : J \to \mathbb{R}_+$ such that for any nonempty bounded set $D \subset E$ we have :

$$\alpha(f(t,D)) \leq \eta(t)\varphi(\alpha(D)) \text{ for a.e. } t \in J.$$

Additionally we assume that

$$\lim_{n\to+\infty} (\psi + \phi)^n(t) = 0 \text{ for a.e } t \in J.$$

(H_5) There exists a positive constant R such that

$$\tilde{M}|y_0| + M|y_1| + \frac{M\psi(R)\|p\|_{L^q}}{\delta^{1-\frac{1}{q}}} \leq R.$$

Theorem 10. *Assume that hypotheses $(H_1) - (H_5)$ are satisfied. If*

$$M \max\left(4\|\eta\|_{L^1}, \frac{\|p\|_{L^q}}{\delta^{1-\frac{1}{q}}}\right) < 1,$$

then the problem (1.1)-(1.2) has a unique S-asymptotically ω-periodic mild solution on J.

Proof. Consider the operator $N : SAP_\omega(X) \to SAP_\omega(X)$ defined by

$$(Ny)(t) = -\frac{\partial}{\partial s}U(t,0)y_0 + U(t,0)y_1 + \int_0^t U(t,s)f(s,y(s))ds. \quad (3.2)$$

Let the operators $\overline{N} = N_1 + N_2$, where

$$(N_1 y)(t) = -\frac{\partial}{\partial s}U(t,0)y_0,$$

$$(N_2 y)(t) = U(t,0)y_1,$$

$$(N_3 y)(t) = \int_0^t U(t,s)f(s,y(s))ds.$$

Claim 1. $\overline{N} \in SAP_\omega(X)$.
For $t \geq 0$, one has

$$\begin{aligned}
&|(\overline{N}y)(t+\omega) - (\overline{N}y)(t)| \\
&= |\tfrac{\partial}{\partial s}U(t+\omega,0)y_0 - \tfrac{\partial}{\partial s}U(t,0)y_0 + U(t+\omega,0)y_1 - U(t,0)y_1| \\
&\leq |\tfrac{\partial}{\partial s}U(t+\omega,0)y_0 - \tfrac{\partial}{\partial s}U(t,0)y_0| + |U(t+\omega,0)y_1 - U(t,0)y_1| \\
&\leq M(e^{-\delta(t+\omega)} - e^{-\delta t})|y_0| + M(e^{-\delta(t+\omega)} - e^{-\delta t})|y_1|.
\end{aligned} \quad (3.3)$$

Since $\delta > 0$, we deduce that

$$\lim_{t\to+\infty} |(\overline{N}y)(t+\omega) - (\overline{N}y)(t)| = 0.$$

Then $\overline{N} \in SAP_\omega(X)$.

Claim 2. $N_3 \in SAP_\omega(X)$.
For $t \geq 0$, one has

$$\begin{aligned}
&(N_3y)(t+\omega) - (N_3y)(t) \\
&= \int_0^{t+\omega} U(t+\omega,s)f(s,y(s))ds - \int_0^t U(t,s)f(s,y(s))ds \\
&= \int_0^{\omega} U(t+\omega,s)f(s,y(s))ds \\
&+ \int_\omega^{t+\omega} U(t+\omega,s)f(s,y(s))ds - \int_0^t U(t,s)f(s,y(s))ds \\
&= \Upsilon_1(t) + \Upsilon_2(t),
\end{aligned}$$

where

$$\Upsilon_1(t) = \int_0^{\omega} U(t+\omega,s)f(s,y(s)ds,$$

and

$$\Upsilon_2(t) = \int_\omega^{t+\omega} U(t+\omega,s)f(s,y(s))ds - \int_0^t U(t,s)f(s,y(s))ds.$$

By (H3), we have

$$\sup_{s\in[0,\omega]} |f(s,y(s))| = f_0 < +\infty.$$

Using the fact that $(U(t,s))_{t\geq s}$ is exponentially stable, we obtain

$$\begin{aligned}
|\Upsilon_1(t)| &\leq Mf_0 \int_0^{\omega} e^{-\delta(t+\omega-s)}ds \\
&= Mf_0e^{-\delta t} \int_0^{\omega} e^{-\delta(\omega-s)}ds \\
&= Mf_0e^{-\delta t}\frac{1-e^{-\delta\omega}}{\delta},
\end{aligned}$$

which shows that

$$\lim_{t\to+\infty} \Upsilon_1(t) = 0.$$

Let us write

$$\Upsilon_2(t) = \int_0^t U(t+\omega, s+\omega)f(s+\omega, y(s+\omega))ds - \int_0^t U(t,s)f(s,y(s))ds$$

and since the evolution family is ω-periodic, we obtain

$$\begin{aligned}\Upsilon_2(t) &\leq \int_0^t U(t,s)\,|f(s+\omega, y(s+\omega)) - f(s+\omega, y(s))|\,ds \\ &+ \int_0^t U(t,s)\,|f(s+\omega, y(s)) - f(s, y(s))|\,ds \\ &= I_1(t) + I_2(t).\end{aligned}$$

Observing that

$$\begin{aligned}|I_1(t)| &\leq \left|\int_0^{T_\varepsilon} U(t,s)\,|f(s+\omega, y(s+\omega)) - f(s+\omega, y(s))|\,ds\right| \\ &+ \left|\int_{T_\varepsilon}^t U(t,s)(f(s+\omega, y(s+\omega)) - f(s+\omega, y(s))ds\right|.\end{aligned}$$

Since $y \in SAP_\omega$, we know that there is a positive constant $T_\varepsilon > 0$ such that

$$|y(t+\omega) - y(t)| \leq \varepsilon \quad \text{for } t \geq T_\varepsilon.$$

By (H_3), we have

$$|f(t, y(s+\omega)) - f(t, y(s))| \leq \frac{\delta\varepsilon}{M} \quad \text{for } t \geq T_\varepsilon.$$

Thus, we get

$$\begin{aligned}&\left|\int_{T_\varepsilon}^t U(t,s)(f(s+\omega, y(s+\omega)) - f(s+\omega, y(s))ds\right| \\ &\leq M\int_{T_\varepsilon}^t e^{-\delta(t-s))}\,|f(s+\omega, y(s+\omega)) - f(s+\omega, y(s))|\,ds \\ &\leq \int_{T_\varepsilon}^t Me^{-\delta(t-s)}\frac{\delta\varepsilon}{M}ds \\ &\leq \varepsilon(e^{-\delta t} - e^{-\delta(t-T_\varepsilon)}) \\ &\leq \varepsilon.\end{aligned} \tag{3.4}$$

Also, we have

$$\begin{aligned}
&\left|\int_0^{T_\varepsilon} U(t,s)(f(s+\omega, y(s+\omega)) - f(s+\omega, y(s))ds\right| \\
&\le M\int_0^{T_\varepsilon} e^{-\delta(t-s)}(|f(s+\omega, y(s+\omega)) - f(s+\omega. y(s))|)ds \\
&\le M\int_0^{T_\varepsilon} e^{-\delta(t-s)}\left(|f(s+\omega, y(s+\omega))| + |f(s+\omega, y(s))|\right)ds \\
&\le 2M\int_0^{T_\varepsilon} e^{-\delta(t-s)}p(s)\psi(|y(s)|)ds \\
&\le 2M\psi(\|y\|_{BC})\int_0^{T_\varepsilon} e^{-\delta(t-s)}p(s)ds.
\end{aligned}$$

For $t \ge 0$, it follows from the Hölder inequality that

$$\begin{aligned}
&\left|\int_0^{T_\varepsilon} U(t,s)(f(s+\omega, y(s+\omega)) - f(s+\omega, y(s))ds\right| \\
&\le 2M\|p\|_{L^q}\psi(\|y\|_{BC})\left[\int_0^{T_\varepsilon} e^{-\frac{q\delta}{q-1}(t-\tau)}d\tau\right]^{1-\frac{1}{q}} \\
&\le \frac{2M\|p\|_{L^q}\psi(\|y\|_{BC})}{\delta^{1-\frac{1}{q}}}\left(e^{-\frac{q\delta}{q-1}t} - e^{-\frac{q\varepsilon}{q-1}(t-T_\varepsilon)}\right)^{1-\frac{1}{q}} \\
&\le \varepsilon.
\end{aligned} \tag{3.5}$$

By (3.4), (3.5) which implies that $\lim_{t\to\infty} I_2(t) = 0$.

By $(H_3)(a)$, we can see that, for every $\varepsilon > 0$, there is a positive constant $T_\varepsilon > 0$ such that

$$|f(t+\omega, y(t)) - f(t, y(t))| \le \varepsilon \quad \text{for } t \ge T_\varepsilon.$$

Similarly, we find

$$\begin{aligned}
&I_2(t)\\
&\leq M\int_{T_\varepsilon}^{t} e^{-\delta(t-s))}\left|f(s+\omega,y(s+\omega))-f(s,y(s))\right|ds\\
&+M\int_0^{T_\varepsilon} e^{-\delta(t-s))}\left|f(s+\omega,y(s))-f(s,y(s))\right|ds\\
&\leq \int_{T_\varepsilon}^{t} Me^{-\delta(t-s)}\frac{\delta\varepsilon}{M}ds\\
&+M\int_0^{T_\varepsilon} e^{-\delta(t-s)}\left(\left|f(s+\omega,y(s))\right|+\left|f(s,y(s))\right|\right)ds\\
&\leq \varepsilon(1-e^{-\delta(t-T_\varepsilon)})\\
&+2M\int_0^{T_\varepsilon} e^{-\delta(t-s)}p(s)\psi(|y(s)|)ds\\
&\leq \varepsilon(1-e^{-\delta(t-T_\varepsilon)})\\
&+2M\psi(\|y\|_\infty)\int_0^{T_\varepsilon} e^{-\delta(t-s)}p(s)ds.
\end{aligned}$$

For $t \geq 0$, it follows from the Hölder inequality that

$$\begin{aligned}
&I_2(t)\\
&\leq \varepsilon(1-e^{-\delta(t-T_\varepsilon)})\\
&+2M\|p\|_{L^q}\psi(\|y\|_\infty)\left[\int_0^{T_\varepsilon} e^{-\frac{q\delta}{q-1}(t-\tau)}d\tau\right]^{1-\frac{1}{q}}\\
&\leq \varepsilon(1-e^{-\delta(t-T_\varepsilon)})\\
&+2\varepsilon\frac{M\|p\|_{L^q}\psi(\|y\|_\infty)}{\delta^{1-\frac{1}{q}}}\left(e^{-\frac{q\delta}{q-1}(t-T_\varepsilon)}-e^{-\frac{q\delta}{q-1}t}\right)^{1-\frac{1}{q}}\\
&\leq \varepsilon.
\end{aligned}$$

Consequently, $\lim_{t\to\infty} I_2(t) = 0$, we conclude that $N_2 \in SAP_\omega$.

From **Claim 1, 2** we deduce $N \in SAP_\omega(E)$.

Next, we will prove that the operator N satisfies all the assumptions of Theorem 8. We will break the proof into several steps.
Let

$$B_R = \{y \in SAP_\omega(E) : \|y\|_\infty \leq R\},$$

where R be any positive constant. Then B_R is a bounded, closed and convex subset of $SAP_\omega(E)$.

Step 1. $N(y) \in B_R$ for any $y \in B_R$.
Now, we have

$$\begin{aligned} |Ny(t)| &\leq \left\| \frac{\partial}{\partial s} U(t,0) \right\|_{B(E)} |y_0| \\ &+ \|U(t,s)\|_{B(E)} |y_1| + \int_0^t \|U(t,s)\|_{B(E)} p(s)\psi(|y(s)|)ds \\ &\leq \tilde{M}|y_1| + M|y_0| + M\psi(R) \int_0^t e^{-\delta(t-s)} p(s)ds. \end{aligned}$$

For $t \geq 0$, it follows from the Hölder inequality that

$$\begin{aligned} |Ny(t)| &\leq \tilde{M}|y_0| + M|y_1| + M\psi(R)\|p\|_{L^q}(1 - e^{-\frac{q\delta}{q-1}t})^{1-\frac{1}{q}} \\ &\leq \tilde{M}|y_0| + M|y_1| + \frac{M\psi(R)\|p\|_{L^q}}{\delta^{1-\frac{1}{q}}} \\ &\leq R. \end{aligned} \tag{3.6}$$

The conditions (H_5) ensure that the operator N transforms the set B_R into itself.

Step 2. N is continuous.
Let $(y_n)_{n\in N}$ be a sequence in B_R such that $y_n \to y$ in B_R.

Case 1. If $t \in [0,T]$; $T > 0$, then, we have

$$\begin{aligned} &|(Ny_n)(t) - (Ny)(t)| \\ &\leq M \int_0^t |f(s, y_n(s)) - f(s, y(s))| \, ds. \end{aligned}$$

Since the function f is Carathéodory, the Lebesgue dominated convergence theorem implies that

$$\|Ny_n - Ny\|_\infty \to 0 \quad \text{as } n \to +\infty.$$

Case 2. Since the function f is Carathéodory, we can see that

$$|f(s, y_n(s)) - f(s, y(s)| \leq \frac{\delta\varepsilon}{M} \quad \text{for } t \geq T. \tag{3.7}$$

If $t \in (T, \infty), T > 0$, then (3.7) and our hypotheses gives us that

$$\begin{aligned}
&|Ny_n(t) - Ny(t)| \\
&\leq \int_0^t \|U(t,s)\|_{B(E)} \Big| f(s, y_n(s)) - f(s, y(s)) \Big| ds \\
&\leq M \frac{\delta\varepsilon}{M} \int_0^t e^{-\delta(t-s)} ds \\
&\leq \frac{M}{\delta} \frac{\delta\varepsilon}{M} (1 - e^{-\delta t}) \\
&\leq \varepsilon.
\end{aligned} \tag{3.8}$$

Then the inequality (3.8) reduces to

$$\|N(u_n) - N(u)\|_\infty \to 0 \quad \text{as } n \to \infty.$$

Now, we conclude that N is continuous from B_R to B_R.

Step 3. $N(B_R)$ is equicontinuous.
Let $t_1, t_2 \in [0, T]$ with $t_2 > t_1$ and $y \in B_R$. Then, we have

$$\begin{aligned}
&|(Ny)(t_2) - (Ny)(t_1)| \\
&= \Bigg| \int_0^{t_1} (U(t_2, s) - U(t_1, s)) f(s, y(s)) \\
&+ \int_{t_1}^{t_2} U(t_2, s) f(s, y(s)) ds \Bigg| \\
&\leq \int_0^{t_1} \|U(t_2, s) - U(t_1, s)\|_{B(E)}\, p(\tau)\psi(|y(s)|) ds \\
&+ M \int_{t_1}^{t_2} e^{-\delta(t-s)} p(s)\psi(|y(s)|) ds. \\
&\leq \int_0^{t_1} \|U(t_2, s) - U(t_1, s)\|_{B(E)}\, p(s)\psi(|y(s)|) ds \\
&+ \frac{M\|p\|_{L^q}\psi(R)}{\delta^{1-\frac{1}{q}}} \left(e^{-\frac{q\delta}{q-1}(t-t_2)} - e^{-\frac{q\delta}{q-1}(t-t_2)} \right)^{1-\frac{1}{q}}.
\end{aligned}$$

The right-hand side of the above inequality tends to zero as $t_2 - t_1 \to 0$, which implies that $N(B_R)$ is equicontinuous.

Consider the measure of noncompactness $\mu(B)$ defined on the family of bounded subsets B of the space $BC(J, E)$ (see [31]) by

$$\mu(B) = \omega_0^T(B) + \sup_{t \in J} \alpha(B(t)) + \lim_{T \to +\infty} \sup\{|y(t)| : t \geq T, y \in B\}.$$

Step 4. $\mu(N(B)) \le M \max\left(4\|\eta\|_{L^1}, \frac{\|p\|_{L^q}}{\delta^{1-\frac{1}{q}}}\right)(\varphi+\psi)(\mu(B))$ for all $B \subset B_R$.
For all $B \subset B_R$, $N(B)$ is bounded. Hence, by Lemma 4, there exists a countable set $B_1 = \{y\}_{n=1}^{\infty} \subset B$, such that

$$\alpha(N(B)) \le 2\alpha(N(B_1)). \tag{3.9}$$

Using the properties of α, Lemma 4, Lemma 5 and assumptions (H_1) and (H_4), we get

$$\begin{aligned}
\alpha(NB_1(t)) &\le \alpha\left(\left\{\int_0^t U(t,s)f(s,y_n(s))ds\right\}_{n=0}^{\infty}\right)\\
&\le 2M\int_0^t \{\alpha\left(f(s,y_n(s))ds)\right)\}_{n=0}^{\infty} ds\\
&\le 2M\int_0^t \eta(s)\varphi\left(\{(\alpha(y_n(s))\}_{n=0}^{\infty})\right)) ds\\
&\le 2M\int_0^t \eta(s)\varphi(\alpha(B(s)))ds.
\end{aligned}$$

Form inequality (3.9), it follows that

$$\alpha(NB(t)) \le 4M\int_0^t \eta(s)\varphi(\alpha(B(s)))ds,$$

then

$$\alpha(N(B(t)) \le 4M\|\eta\|_{L^1}\varphi(\sup_{t\in J}\alpha(B(t))).$$

Since

$$\sup_{t\in J}\alpha(B(t)) \le \sup_{t\in J}\alpha(B(t)) + \lim_{T\to+\infty}\sup\{|y(t)| : t \ge T, y \in B\},$$

then

$$\alpha(N(B(t)) \le 4M\|\eta\|_{L^1}\varphi(\sup_{t\in J}\alpha(B(t)) + \lim_{T\to+\infty}\sup\{|y(t)| : t \ge T, y \in B\}). \tag{3.10}$$

On the other hand, for $t \geq T$, we have

$$\begin{aligned}
|(Ny)(t)| &\leq \tilde{M}e^{-\delta t}|y_1| + Me^{-\delta t}|y_0| \\
&+ M\int_0^t e^{-\delta(t-s)}p(s)\psi(|y(s)|)ds. \\
&\leq \tilde{M}e^{-\delta t}|y_1| + Me^{-\delta t}|y_0| \\
&+ M\int_0^T e^{-\delta(t-s)}p(s)\psi(|y(s)|)ds. \\
&+ M\int_T^t e^{-\delta(t-s)}p(s)\psi(|y(s)|)ds. \\
&\leq \tilde{M}e^{-\delta t}|y_1| + Me^{-\delta t}|y_0| \\
&+ M\int_0^T e^{-\delta(t-s)}p(s)ds\psi(\sup_{s\in[0,T]}|y(s)|). \\
&+ M\int_T^t e^{-\delta(t-s)}p(s)ds\psi(\sup\{|y(t)| : t \geq T, y \in B\}).
\end{aligned}$$

Next, applying the Hölder inequality we derive

$$\begin{aligned}
|(Ny)(t)| &\leq (\tilde{M}e^{-\delta t}|y_1| + Me^{-\delta t}|y_0|) \\
&+ \frac{M\|p\|_{L^q}}{\delta^{1-\frac{1}{q}}}(e^{-\frac{q\delta}{q-1}t} - e^{-\frac{q\delta}{q-1}(t-T)})^{1-\frac{1}{q}}\psi(\|y\|_\infty). \\
&+ \frac{M\|p\|_{L^q}}{\delta^{1-\frac{1}{q}}}(1 - e^{-\frac{q\delta}{q-1}t})^{1-\frac{1}{q}}\psi(\sup\{|y(t)| : t \geq T, y \in B\}).
\end{aligned}$$

Then

$$\begin{aligned}
|(Ny)(t)| &\leq (\tilde{M}e^{-\delta t}|y_1| + Me^{-\delta t}|y_0|) \\
&+ \frac{M\|p\|_{L^q}}{\delta^{1-\frac{1}{q}}}e^{-\delta T}\psi(\|y\|_\infty). \\
&+ \frac{M\|p\|_{L^q}}{\delta^{1-\frac{1}{q}}}\psi(\sup\{|y(t)| : t \geq T, y \in B\}).
\end{aligned}$$

Since

$$\lim_{T\to+\infty}\sup\{|y(t)| : t \geq T, y \in E\} \leq \sup_{t\in J}\alpha(B(t)) + \lim_{T\to+\infty}\sup\{|y(t)| : t \geq T, y \in B\},$$

then

$$\begin{aligned}&\lim_{T\to+\infty}\sup\{|(Ny)(t):t\geq T,y\in B\})\\&\leq\frac{M\|p\|_{L^q}}{\delta^{1-\frac{1}{q}}}\psi(\sup_{t\in J}\alpha(B(t))+\lim_{T\to+\infty}\sup\{|y(t)|:t\geq T,y\in B\}).\end{aligned}\tag{3.11}$$

Further, combining (3.10) and (3.11), we get

$$\begin{aligned}&\sup_{t\in J}\alpha((NB)(t))+\lim_{T\to+\infty}\sup\{|(Ny)(t):t\geq T,y\in B\}\\&\leq 4M\|\eta\|_{L^1}\varphi(\sup_{t\in J}\alpha(B(t))+\lim_{T\to+\infty}\sup\{|y(t)|:t\geq T,y\in B\})\\&+\frac{M\|p\|_{L^q}}{\delta^{1-\frac{1}{q}}}\psi(\sup_{t\in J}\alpha(B(t))+\lim_{T\to+\infty}\sup\{|y(t)|:t\geq T,y\in B\}\\&\leq M\max(4\|\eta\|_{L^1},\frac{\|p\|_{L^q}}{\delta^{1-\frac{1}{q}}})(\varphi+\psi)(\sup_{t\in J}\alpha(B(t))+\lim_{T\to+\infty}\sup\{|y(t)|:t\geq T,y\in B\}).\end{aligned}\tag{3.12}$$

From **Step 3** and inequality (3.12), we conclude that

$$\mu(N(B))\leq M\max\left(4\|\eta\|_{L^1},\frac{\|p\|_{L^q}}{\delta^{1-\frac{1}{q}}}\right)(\varphi+\psi)(\mu(B)).$$

It follows from Lemma 8 that N has at least one fixed point $y\in B_R$, which is mild solution of problem (1.1)-(1.2) on the interval J.

4. An Example

Consider the second order differential equation of the form;

$$\left\{\begin{array}{rll}\frac{\partial^2}{\partial t^2}z(t,\tau) & =\frac{\partial^2}{\partial\tau^2}z(t,\tau)+(-3+\cos(\pi t))\frac{\partial}{\partial t}z(t,\tau) & \\ & +\frac{t^{\frac{1}{r}}}{2(1+t^2)^{\frac{2}{r}}}\ln\left(1+\frac{2t^{\frac{r-1}{r}}}{(1+t^2)^{\frac{2r-2}{r}}}|z(t,\tau)|\right) & \\ & r>1,\ t\in J,\ \tau\in[0,\pi], & \\ z(t,0)= & z(t,\pi)=0 & t\in J,\\ z(t+1,\tau)= & z(t,\tau) & t\in J,\ \tau\in[0,\pi],\\ \frac{\partial}{\partial t}z(0,\tau) & =\psi(\tau) & \tau\in[0,\pi]\end{array}\right.\tag{4.1}$$

Let $E=L^2([0,\pi],\mathbb{R})$ be the space of 2-integrable functions from $[0,\pi]$ into $\mathbb{R}$, and let $H^2([0,\pi],\mathbb{R})$ be the Sobolev space of functions $x:[0,\pi]\to\mathbb{R}$, such that $x''\in L^2([0,\pi],\mathbb{R})$. We consider the operator $A_1z(\tau)=z''(\tau)$ with domain $D(A_1)=H^2([0,\pi],\mathbb{C})$, which is the infinitesimal generator of strongly continuous cosine function $C(t)$ on E. Moreover, A_1 has discrete spectrum, the spectrum of A_1 consists of eigenvalues n^2 for $n \in \mathbb{Z}$, with associated eigenvector

$$\omega_n(\xi) = \frac{1}{\sqrt{2\pi}} e^{in\xi}, n \in \mathbb{Z},$$

the set $\{\omega_n, n \in \mathbb{Z}\}$ is an orthonormal basis of E. In particular,

$$A_1 x = -\sum_{n=1}^{\infty} n^2 \langle x, w_n \rangle w_n \text{ for } x \in D(A).$$

The cosine function $C(t)$ is given by

$$C(t)x = \sum_{n=1}^{\infty} \cos(nt) \langle x, w_n \rangle w_n \text{ for } x \in D(A), t \in \mathbb{R}.$$

The associated sine function is defined by

$$S(t)x = \sum_{n=1}^{\infty} \frac{\sin(nt)}{n} \langle x, w_n \rangle w_n \text{ for } x \in D(A), t \in \mathbb{R}.$$

From [34], for all $x \in H^2([0,\pi], \mathbb{R}), t \in \mathbb{R}$,

$$\|C(t)\|_{B(E)} \leq e^{-t}, \text{ and } \|S(t)\|_{B(E)} \leq e^{-t}.$$

Now, we define an operator $A(t) : D(A) \subset L^2([0,\pi], \mathbb{C}) \to L^2([0,\pi], \mathbb{C})$ by

$$\begin{cases} D(A(t)) = D(A) \\ A(t) = A_1 + b(t,\tau), \end{cases} \quad (4.2$$

where $b(t,\tau) = -3 + 2\cos \pi t$.
Note that $A(t)$ generates an evolution system $U(t,s)$ of the form

$$U(t,s) = S(t-s) e^{\int_s^t b(t,s)ds}.$$

Since $b(t,\tau) = -3 + \cos(\pi t) \leq 2$, we have

$$U(t,s) = S(t-s) e^{-2(t-s)} \quad (4.3$$

and

$$\|U(t-s)\|_{B(E)} \leq \|S(t-s)\|_{B(E)} e^{-2(t-s)} \leq e^{-3(t-s)}.$$

Since $b(t+2,x) = b(t,x)$, we conclude that $U(t,s)$ is a 2-periodic evolution system exponentially stable. It follows from the estimate (4.3) that $U(t,s)$ is well defined and satisfies the conditions of Definition 1. Hence conditions $(H1)$ and $(H2)$ are satisfied.

Set

$$z(t)(\tau) = w(t)(\tau),\ t \geq 0,\ \tau \in [0,\pi],$$

$$f(t,u)(\tau) = \frac{t^{\frac{1}{r}}}{2(1+t^2)^{\frac{2}{r}}} \ln\left(1 + \frac{2t^{\frac{r-1}{r}}}{(1+t^2)^{\frac{2r-2}{r}}}|z(,\tau)|\right).$$

Further, for arbitrarily fixed $z \in H^2([0,\pi],\mathbb{R})$ and for $t > 0$ we obtain

$$\begin{aligned}
&|f(t+1,z)(\tau) - f(t,z)(\tau)| \\
&\leq \frac{t^{\frac{1}{r}}}{2(1+t^2)^{\frac{2}{r}}} \ln\left(1 + \frac{2t^{\frac{r-1}{r}}}{(1+t^2)^{\frac{2r-2}{r}}}|z(t+1,\tau)|\right) \\
&- \frac{t^{\frac{1}{r}}}{2(1+t^2)^{\frac{2}{r}}} \ln\left(1 + \frac{2t^{\frac{r-1}{r}}}{(1+t^2)^{\frac{2r-2}{r}}}|z(t,\tau)|\right) \\
&\leq \frac{t^{\frac{1}{r}}}{2(1+t^2)^{\frac{2}{r}}} \ln\left(\frac{1+2(t+1)^{\frac{r-1}{r}}|z(t+1,\tau)|}{1+2t^{\frac{r-1}{r}}|z(t,\tau)|}\right) \\
&\leq \frac{t^{\frac{1}{r}}}{2(1+t^2)^{\frac{2}{r}}} \ln\left(\left(\frac{t+1}{t}\right)^{\frac{r-1}{r}} + \frac{1-\left(\frac{t+1}{t}\right)^{\frac{r-1}{r}}}{1+2t^{\frac{r-1}{r}}|z(t,\tau)|}\right) \\
&\leq \frac{t^{\frac{1}{r}}}{2(1+t^2)^{\frac{2}{r}}} \ln\left(\left(\frac{t+1}{t}\right)^{\frac{r-1}{r}} + 1\right).
\end{aligned} \tag{4.4}$$

Now from (4.4), we can see $|f(t+1,z)(\tau) - f(t,z)(\tau)|$ tends to zero as $t \to \infty$. Hence condition $(H3)(a)$ is satisfied with $\omega = 1$.

We can estimate the function f:

$$f(t,u)(\tau) \leq \frac{t^{\frac{1}{r}}}{2(1+t^2)^{\frac{2}{r}}} \ln(1 + 2|z(t,\tau)|).$$

Hence condition $(H3)(b)$ is satisfied with

$$p(t) = \frac{t^{\frac{1}{r}}}{2(1+t^2)^{\frac{2}{r}}}, \quad \psi(t) = \frac{1}{4}\ln(1+2t).$$

We can also estimate the function f:

$$f(t,u)(\tau) \leq \frac{t}{(t^2+1)^2}|z(t,\tau)|. \qquad (4.5)$$

By (4.5), for every $t \in J$, and $B \in D \subset E$, we have

$$\alpha(f(t,D) \leq \frac{t}{2(1+t^2)^2}\alpha(D),$$

Hence condition $(H4)$ is satisfied with

$$\eta(t) = \frac{t}{(1+t^2)^2}, \quad \varphi(t) = \frac{t}{2}.$$

Moreover, we have

$$(\psi+\varphi)(t) = \frac{t}{2} + \frac{1}{4}\ln(1+2t) \leq t.$$

We conclude that (see Lemma 2.1. [6])

$$\lim_{n \to +\infty} (\psi+\phi)^n(t) = 0 \text{ for a.e } t \in J.$$

Consequently, (4.1) can be written in the abstract form (1.1)-(1.2) with $A(t$ and f as defined above. The existence of mild solutions can be deduced fron an application of Theorem 10.

CONCLUSION

In this article, we study sufficient conditions for the existence of S asymptotically ω-periodic mild solution for second differential equation of problem (1.1)-(1.2) assuming that $f(\cdot,\cdot)$ is S-asymptotically ω-periodi in the first variable and not Lipschitz in the second variable. The argument are based upon on a new fixed point result which is a generalization of th well known Darbo's fixed point theorem and the technique of measures o noncompactness.

REFERENCES

[1] Abbas S. and Benchohra M., *Advanced Functional Evolution Equations and Inclusions*, Springer, Cham, 2015.

[2] Ahmed N. U., *Semigroup Theory with Applications to Systems and Control*, Harlow John Wiley & Sons, Inc., New York, 1991.

[3] Aissani K., Benchohra M., Global existence results for fractional integro-differential equations with state-dependent delay. *An. tiin. Univ. Al. I. Cuza Iai. Mat. (N.S.)* 62 (2016), 411-422.

[4] Akhmerov R. R., Kamenskii M. I., Patapov A. S., Rodkina A. E. and Sadovskii B. N., *Measures of Noncompactness an Condensing Operators*, Birkhauser Verlag, Basel, 1992.

[5] Alvárez J. C., Measure of Noncompactness and fixed points of nonexpansive condensing mappings in locally convex spaces, *Rev. Real. Acad. Cienc. Exact. Fis. Natur. Madrid* **79** (1985), 53-66.

[6] Aghajani A., Banaŝ J., Sabzali N., Some generalizations of Darbo fixed point theorem and applications. *Bull. Belg. Math. Soc. Simon Stevin* 20 (2013), no. 2, 345-358.

[7] Balachandran K., Park D. G. and Anthoni S. M., Existence of solutions of abstract nonlinear second-order neutral functional integrodifferential equations, *Comput. Math. Appl.* **46** (2003), 1313-1324.

[8] Banaŝ J. and Goebel K., Measures of Noncompactness in Banach Spaces. *Lecture Note in Pure App. Math.* 60, Dekker, New York, 1980.

[9] Batty C. J. K., Chill R., Srivastava S., Maximal regularity for second order non-autonomous Cauchy problems, *Studia Math.* 189 (2008) 205-223.

[10] Benchohra M, Henderson J. and Rezoug N., Global existence results for second order evolution equations, *Comm. Appl. Nonlinear Anal.* 23 (2016), 57-67.

[11] Blot J., Cieutat P., N'Guérékata G. M., S-asymptotically ω-periodic functions applications to evolution equations, *African Diaspora J. Math.*, 12 (2011), 113-121.

[12] Chen P., Li Y., Monotone iterative technique for a class of semilinear evolution equations with nonlocal conditions, *Results Math.* 63 (2013), 731-744.

[13] Cuevas C. and Souza J. C., S-asymptotically ω-periodic solutions of semilinear fractional integro-differential equations. *Appl. Math. Lett.* 22 (2009), 865-870.

[14] Cuevas C. and Lizanna C., Existence of S-asymptotically ω-periodic solutions for two-times fractional order differential equations. *Southeast Asian Bull. Math.* 37 (2013), no. 5, 683-690.

[15] Cuevas C. and Lizanna C., S-asymptotically ω-periodic solutions of semilinear Volterra equations. *Math. Meth. Appl. Sci.* 33 (2010), 1628-1636.

[16] Diagana T., Almost periodic solutions to some second-order nonautonomous differential equations. *Proc. Am. Math. Soc.* 140 (2012), 279-289.

[17] Dimbour W., Mado J-C., S-asymptotically ω-periodic solution for a nonlinear differential equation with piecewise constant argument in a Banach space. *Cubo* 16 (2014), no. 3, 55-65.

[18] Dimbour W., N'Guérékata G. M., S-asymptotically ω-periodic solutions to some classes of partial evolution equations. *Appl. Math. Comput.* 218 (2012), 7622-7628.

[19] Faraci F., Iannizzotto A, A multiplicity theorem for a perturbed second-order non-autonomous system, *Proc. Edinb. Math. Soc.* 49 (2006) 267-275.

[20] Fattorini H. O., *Second Order Linear Differential Equations in Banach Spaces*, North-Holland Mathematics Studies, Vol. **108**, North-Holland Amsterdam, 1985.

[21] Henríquez H. R., Pierri M., Taboas P., Existence of S-asymptotically ω-periodic solutions for abstract neutral equations, *Bull. Aust. Math. Soc.*, 7[illegible] (2008), 365-382.

[22] Henríquez H. R., Pierri M., Taboas P., On S-asymptotically ω-periodic functions on Banach spaces and applications, *J. Math. Anal. Appl.*, 34[illegible] (2008), 1119-1130.

[23] Henríquez H., Poblete V, Pozo J., Mild solutions of non-autonomous second order problems with nonlocal initial conditions. *J. Math. Anal. Appl.* 412 (2014), no. 2, 1064-1083.

[24] Heinz H. P., On the behaviour of measure of noncompactness with respect to differentiation and integration of rector-valued functions, *Nonlinear Anal.* 7 (1983), 1351-1371.

[25] Kozak M., A fundamental solution of a second-order differential equation in a Banach space, *Univ. Iagel. Acta Math.* **32** (1995), 275-289.

[26] Mönch H., Boundry value problems for nonlinear ordinary differential equations of second order in Banach spaces, *Nonlinear Anal.* 4(5) (1980) 985-999.

[27] N'Guérékata G. M., *Almost Automorphic and Almost Periodic Functions in Abstract Spaces,* Kluwer Academic, New York, London, Moscow, 2001.

[28] Li F., Liang J., Wang H., S-asymptotically ω-periodic solution for fractional differential equations of order $q \in (0, 1)$ with finite delay. *Adv. Difference Equ.* (2017), Paper No. 83, 14 pp.

[29] Li F., Wang H, S-asymptotically ω-periodic mild solutions of neutral fractional differential equations with finite delay in Banach space. *Mediterr. J. Math.* 14 (2017), no. 2, Art. 57, 16 pp.

[30] Liu J. H., Bounded and periodic solutions of semi-linear evolution equations. *Dyn. Syst. Appl.* 4(1995), 341-350.

[31] Olszowy L., Wędrychowicz S., On the existence and asymptotic behaviour of solutions of an evolution equation and an application to the Feynman-Kac theorem. *Nonlinear Anal.* 74 (2011), no. 17, 6758-6769.

[32] Olszowy L., Existence of mild solutions for semilinear nonlocal Cauchy problems in separable Banach spaces. *Z Anal. Anwend.* 32 (2) (2013), 215-232.

[33] Pierri M., On S-Asymptotically -periodic functions and applications. *Nonlinear Anal.* 75 (2012), no. 2, 651–661.

[34] Travis C. C. and Webb G. F., Second order differential equations in Banach spaces, in: Nonlinear Equations in Abstract Spaces, *Proc. Internat. Sympos.* (Univ. Texas, Arlington, TX, 1977), Academic Press, New-York, 1978, 331-361.

[35] Winiarska T., Evolution equations of second order with operator dependent on t, *Selected Problems of Math. Cracow Univ. Tech.* 6 (1995) 299-314.

[36] Wu J., *Theory and Application of Partial Functional Differential Equations,* Springer-Verlag, New York, 1996.

[37] Xie R. and Zhang C., Criteria of asymptotic ω-periodicity and their applications in a class of fractional differential equations. *Adv. Difference Equ.* 2015, 2015:68, 20 pp.

In: Evolutionary Processes and Applications
Editor: G. M. N'Guérékata
ISBN: 978-1-53616-350-6

Chapter 10

THE INITIAL BOUNDARY VALUE PROBLEM OF AN EVOLUTIONARY $p(x)$-LAPLACIAN EQUATION

***Zhao Jiangyan*[1],* *and Huashui Zhan*[2,†]**
[1]Center for General Education, Xiamen Huaxia University, China
[2]School of Applied Mathematics,
Xiamen University of Technology, China

Abstract

Consider the initial boundary value problem of an equation

$$u_t = div(|\nabla u|^{p(x)-2}\nabla u) + f(x,t,u).$$

The problems describe the motion of electrorheological fluids. If the variable exponent $p(x)$ is required to satisfy a logarithmic Hölder continuity condition, there are many references to study the equation. If the exponent $p(x)$ does not satisfy the logarithmic Hölder continuity condition, the related references are much fewer. If $\inf p(x) > 2$, Lian etc. proved the existence and uniqueness of the equation in [1]. In the chapter, by difference method, the case of the equation with $2 > ess\sup p(x) \geq ess\inf p(x) > 1$ is studied. The existence and uniqueness of weak solutions are obtained provided that $ess\inf p(x) > \frac{2N}{N+1}$.

Keywords: electrorheological fluids, $p(x)$-Laplacian equation, weak solution, difference method

*Corresponding Author's Email: zhaojy@hxxy.edu.cn.
†Corresponding Author's Email: 2012111007@xmut.edu.cn.

1. Introduction and Main Results

Let $\Omega \subset R^N$ be a bounded domain with suitably smooth boundary $\partial\Omega$. For the following equation

$$u_t = div(|\nabla u|^{p(x)-2}\nabla u) + f(x,t,u) \tag{1.1}$$

where $p(x)$ is a measurable function, $x \in \Omega$, $0 < t < T$. If we want to consider its initial boundary value problem, the initial value condition

$$u|_{t=0} = u_0 \tag{1.2}$$

is always required, and due to $ess\,\inf p(x) > 1$, according to Oleinik rules [2], we also can endowed the general homogeneous boundary value condition

$$u|_{\Gamma_T} = 0, \tag{1.3}$$

where $\Gamma_T = \partial\Omega \times (0,T]$.

In the case when p is a constant, the problem describes the motion of generalized non-Newtonian flow, there are many works about the existence, uniqueness and the regularity of the solutions. The bibliography is given in [3, 4, 5] and the references therein. The second author also have published a series papers on this problem, see [18]-[28].

If $p(x)$ is only a measurable function, a new interesting kind of fluids of prominent technological interest has recently emerged: the so-called electrorheological fluids. This model includes parabolic equations which are nonlinear with respect to the gradient of the thought solution, and with variable exponents of nonlinearity. The typical case is the so-called evolutionary $p(x)$-Laplacian equation with exponent p as a function of the external electromagnetic field (see [6, 7] and the references therein).

Just as Sobolev space $W^{1,p}(\Omega)$ is used to study the evolutionary $p-Laplacian$ equation, we first introduce the functional space $W^{1,p(x)}(\Omega)$, for more details, see [11, 12].

1. $L^{p(x)}(\Omega)$ space.

$$L^{p(x)}(\Omega) = \{u : u \ is\ a\ measurable\ real-valued\ function, \int_\Omega |u(x)|^{p(x)}dx < \infty\}.$$

The space $L^{p(x)}(\Omega)$ is equipped with the following Luxemburg's norm

$$|u|_{L^{p(x)}(\Omega)} = \inf\{\lambda > 0 : \int_\Omega |\frac{u(x)}{\lambda}|^{p(x)}dx \le 1\}.$$

The space $\left(L^{p(x)}(\Omega), |.|_{L^{p(x)}(\Omega)}\right)$ is a separable ,uniformly convex Banach space.

2. $W^{1,p(x)}(\Omega)$ space.

$$W^{1,p(x)}(\Omega) = \{u \in L^{p(x)}(\Omega) : |\nabla u| \in L^{p(x)}(\Omega)\},$$

endowed with the following norm

$$|u|_{W^{1,p(x)}(\Omega)} = |u|_{L^{p(x)}(\Omega)} + |\nabla u|_{L^{p(x)}(\Omega)}, \forall u \in W^{1,p(x)}(\Omega).$$

We use $W_0^{1,p(x)}(\Omega)$ to denote the closure of $C_0^\infty(\Omega)$ in $W^{1,p(x)}(\Omega)$.

Lemma 1. *(i) The space* $\left(L^{p(x)}(\Omega), |\cdot|_{L^{p(x)}(\Omega)}\right)$, $\left(W^{1,p(x)}(\Omega), |\cdot|_{W^{1,p(x)}(\Omega)}\right)$ *and* $W_0^{1,p(x)}(\Omega)$ *are reflexive Banach spaces.*

(ii) $p(x)$-Hölder's inequality. Let $q_1(x)$ and $q_2(x)$ be real functions with $\frac{1}{q_1(x)} + \frac{1}{q_2(x)} = 1$ and $q_1(x) > 1$. Then, the conjugate space of $L^{q_1(x)}(\Omega)$ is $L^{q_2(x)}(\Omega)$. And for any $u \in L^{q_1(x)}(\Omega)$ and $v \in L^{q_2(x)}(\Omega)$, we have

$$\left|\int_\Omega uv dx\right| \leq 2|u|_{L^{q_1(x)}(\Omega)}|v|_{L^{q_2(x)}(\Omega)}.$$

(iii)

$$|u|_{L^{p(x)}(\Omega)} = 1, then \int_\Omega |u|^{p(x)} dx = 1,$$

$$|u|_{L^{p(x)}(\Omega)} > 1, then |u|^{p^-}_{L^{p(x)}(\Omega)} \leq \int_\Omega |u|^{p(x)} dx \leq |u|^{p^+}_{L^{p(x)}(\Omega)},$$

$$|u|_{L^{p(x)}(\Omega)} < 1, then |u|^{p^+}_{L^{p(x)}(\Omega)} \leq \int_\Omega |u|^{p(x)} dx \leq |u|^{p^-}_{L^{p(x)}(\Omega)}.$$

(iv) If $p_1(x) \leq p_2(x)$, then

$$L^{p_1(x)}(\Omega) \supset L^{p_2(x)}(\Omega).$$

(v) If $p_1(x) \leq p_2(x)$, then

$$W^{1,p_2(x)}(\Omega) \hookrightarrow W^{1,p_1(x)}(\Omega).$$

Besides this trivial embedding, it would be useful to know finer estimates of the type of Sobolev inequality. We may ask whether there exists the embedding

$W^{1,p(x)}(\Omega) \hookrightarrow L^{q(x)}(\Omega)$, with $\frac{1}{q(x)} = \frac{1}{p(x)} - \frac{1}{N}, x \in \Omega$,in general, this we cannot expect.

(vi) $p(x)$-Poincarés inequality. If $p(x) \in C(\overline{\Omega})$, then there is a constant $C > 0$, such that

$$|u|_{L^{p(x)}(\Omega)} \leq C|\nabla u|_{L^{p(x)}(\Omega)}, \quad \forall u \in W_0^{1,p(x)}(\Omega).$$

This implies that $|\nabla u|_{L^{p(x)}(\Omega)}$ and $|u|_{W^{1,p(x)}(\Omega)}$ are equivalent norms of $W_0^{1,p(x)}(\Omega)$.

However, in [8], Zhikov showed that

$$W_0^{1,p(x)}(\Omega) \neq \{v \in W_0^{1,p(x)}(\Omega), v|_{\partial\Omega} = 0\} = \mathring{W}^{1,p(x)}(\Omega).$$

Hence, the property of the space is different from the case when p is a constant. This fact makes the general methods, for examples, the methods used in [3, 4, 5] and [18]-[28], to study the posedness of evolutionary $p-$Laplacian equation cannot be used directly.

If the exponent $p(x)$ is required to satisfy a logarithmic Hölder continuity condition

$$|p(x) - p(y)| \leq \omega(|x - y|), \forall x, y \in Q_T, \quad |x - y| < \frac{1}{2}, \qquad (1.4)$$

with

$$\overline{\lim_{s\to 0^+}}\, \omega(s) \ln(\frac{1}{s}) = C < \infty,$$

then (see [8])

$$W_0^{1,p(x)}(\Omega) = \mathring{W}^{1,p(x)}(\Omega).$$

By the fact, Antontsev-Shmarev [9] established the existence and uniqueness results of equation (1.1).

Throughout the paper we denote

$$p^+ = \operatorname{ess}\sup_{\bar{\Omega}} p(x), p^- = \operatorname{ess}\inf_{\bar{\Omega}} p(x),$$

and assume that

$$\frac{2N}{N+1} < p^- \leq p(x) \leq p^+, \quad \forall x \in \Omega.$$

When $p^- > 2$, Peng [13] had studied the existence of the solutions of the equation

$$\partial_t u - P(u)(\nabla H(u)) - div(|\nabla u|^{p-2}\nabla u) = |\nabla u|^{p-2}A(u)(\nabla u, \nabla u)$$

where $(x,t) \in Q_T = \Omega \times (0,T)$, without the condition (1.4). By adopting a method of difference in time, Lian etc. [1] generalized [13]'s method to study the equation (1.1)

$$u_t = \operatorname{div}(|\nabla u|^{p(x)-2}\nabla u) - f(x,t,u),$$

provided that $p^- > 2$ and f satisfied some restrictions. Some other papers in studying the regularity of evolutionary $p(x)$-Laplacian equation, for examples, one can refer to [14, 15] et.al.

In our paper, we continue to study the initial boundary value problem of equation (1.1)-(1.3) with $2 > p^+ = es\,\mathrm{sup}p(x) \geq ess\,\mathrm{inf}p(x) > \frac{2N}{N+1}$. Though the outline of the proof in our paper is similar as [1] and [13], due to $ess\,\mathrm{sup}p(x) < 2$, some different techniques are used.

The definition of the solution to problem (1.1)-(1.3) is as follows.

Definition 2. *A function u is said to be a weak solution of (1.1)-(1.3), if u satisfies the following:*

$$u \in L^2(Q_T),\ f(x,t,u) \in L^1(Q_T), D_i u \in L^{p(x)}(Q_T),$$

$$u = 0\ on\ \partial\Omega \times (0,T)\ in\ the\ sense\ of\ trace.$$

For all $\varphi \in C_0^\infty(Q_T)$

$$\int_0^T \int_\Omega (u\frac{\partial \varphi}{\partial t} - |\nabla u|^{p(x)-2}\nabla u \cdot \nabla\varphi + f\varphi)dxdt = 0, \tag{1.5}$$

and

$$\lim_{t\to 0}\int_\Omega (u(x,t) - u_0(x))\psi(x)dx = 0, \tag{1.6}$$

for all $\psi \in C_0^\infty(\Omega)$, where $Q_T = \Omega \times (0,T)$.

As the reference [1], in the study of the global existence of solutions, the following hypotheses to the function f are required:

$$f(x,t,z) \in C^1(\bar{\Omega} \times [0,T] \times R)$$

and

$$|f(x,t,z)| \leq C_0(\phi(x,t) + |z|^\alpha), \tag{A}$$

where $\phi \geq 0, \phi \in L^r(\Omega \times (0,T)), r > \frac{N+p^-}{p^-}$ and $C_0 > 0, \alpha \geq 0$ are constants.

Our main results are the following theorems.

Theorem 3. *Let $u_0 \in L^\infty(\Omega) \bigcap W_0^{1,p(x)}(\Omega)$ and (A) hold. Assume that*

$$\alpha < p^- - 1 \text{ or } \quad \alpha = p^- - 1 \text{ and } |\Omega| \text{ is small enough}, \tag{B}$$

where $|\Omega|$ denotes the Lebesgue measure of Ω. If

$$2 > p^+ = ess\sup p(x) \geq ess\inf p(x) > \frac{2N}{N+1},$$

then there exists a weak solution of (1.1)-(1.3) such that

$$u \in L^\infty(Q_T) \cap L^\infty(0,T; W_0^{1,p(x)}(\Omega)), u_t \in L^2(\Omega \times (0,T)).$$

Remark 4. *The condition (B) implies that $\alpha \leq 1$. which is different from the case when $ess\inf p(x) > 2$ (see [1]). The constant α in [1] only is nonnegative. At the same time, by comparing to the case of that p is only a constant, the best condition of $p(x)$ in Theorem 1.3 should be as $2 > p^+ = ess\sup p(x) \geq ess\inf p(x) > \frac{2N}{N+1}$.*

Theorem 5. *If*

$$f(x,t,z) \in C^1(\bar{\Omega} \times [0,T] \times R), 2 > p^+ = ess\sup p(x) \geq ess\inf p(x) > \frac{2N}{N+1},$$

then the solution of (1.1)-(1.3) with

$$u \in L^\infty(Q_T), u_t \in L^2(\Omega \times (0,T)).$$

is unique.

2. The Difference Equation of the Problem

We introduce that

$$F^i(x,u) = \int_{u_{i-1}}^{u} \left(\frac{1}{h} \int_{ih}^{(i+1)h} f(x,\tau,s)d\tau \right) ds, \tag{2.1}$$

and

$$\psi^i(u) = \int_\Omega \frac{1}{p(x)}|\nabla u|^{p(x)}dx - \int_\Omega F^i(x,u)dx + \frac{1}{2h}\int_\Omega (u-u_{i-1})^2 dx, i = 1, 2, ... \tag{2.2}$$

Denote

$$p^* = \begin{cases} \frac{Np^-}{N-p^-}, & \text{if } N > p^- > 1, \\ (\frac{N+p^-}{N})p^-, & \text{if } N \le p^-. \end{cases}$$

one can see that $N \le p^-$ means that $N = 1$, except that $p^- = p(x) = p^+ = 2$, the last case means that $N = 2$, and equation (1.1) becomes the equation of heat conduction, which had been thoroughly studied, and we shall not discuss this case in what follows. So

$$p^* = \begin{cases} \frac{Np^-}{N-p^-}, & \text{if } N > p^- > 1, \\ (1+p^-)p^-, & \text{if } N = 1. \end{cases} \tag{2.3}$$

Lemma 6. *Assume that $p(x) \in C(\bar{\Omega}), u_{i-1}(x) \in L^{p*}(\Omega)$ and (A), (B) hold. Then the functional $\psi^i(u)$ achieves its minimum on the set*

$$S = \{u \in W_0^{1,p(x)}(\Omega)\}. \tag{2.4}$$

Proof. The proof is just as [1]. In the proof, one can see that the conditions $\alpha < 1$ and $p^- > \frac{2N}{N+1}$ can guarantee the method used in [1] is also valid in the case of $p^+ < 2$. As in [1], we shall check that $\psi^i(u)$ satisfies the conditions which assure the existence of a minimum on the set.

1. S is weakly closed.

By Lemma 1.1(i) we know that $W_0^{1,p(x)}(\Omega)$ is a reflexive Banach space and then by Mazur theorem it is weakly closed.

2. $\psi^i(u)$ satisfies the coerciveness conditions. By (A) we have

$$\begin{aligned}\psi^i(u) \ge & \int_\Omega \frac{1}{p(x)}|\nabla u|^{p(x)}dx - C_0\int_\Omega \left(\frac{1}{h}\int_{ih}^{(i+1)h}\phi(x,\tau)d\tau\right)|u-u_{i-1}|dx \\ & - C_1\int_\Omega (|u|^{\alpha+1} + |u_{i-1}|^{\alpha+1})dx + \frac{1}{2h}\int_\Omega (u-u_{i-1})^2 dx,\end{aligned} \tag{2.5}$$

where $C_1 > 0$ is a constant.

We first estimate the second term on the right-hand side of the inequality.
Denote $r_1 = \frac{N+p^-}{p^-}$ and $r_2 = \frac{N+p^-}{N}$.
By (A) and the Hölder inequality, we get

$$\begin{aligned}
I_1 &= C_0 \int_\Omega \left(\frac{1}{h} \int_{ih}^{(i+1)h} \phi(x,\tau) d\tau \right) |u - u_{i-1}| dx \\
&\le C_0 \left(\int_\Omega \left(\frac{1}{h} \int_{ih}^{(i+1)h} \phi(x,\tau) d\tau \right)^{r_1} dx \right)^{\frac{1}{r_1}} \left(\int_\Omega |u - u_{i-1}|^{r_2} dx \right)^{\frac{1}{r_2}} \\
&\le C \left(\int_\Omega \frac{1}{h} \int_{ih}^{(i+1)h} \phi^{r_1}(x,\tau) d\tau dx \right)^{\frac{1}{r_1}} \left(\int_\Omega |u - u_{i-1}|^{r_2} dx \right)^{\frac{1}{r_2}} \\
&\le C \|u - u_{i-1}\|_{L^{r_2}(\Omega)} \le C \|u\|_{L^{r_2}(\Omega)} + \|u_{i-1}\|_{L^{r_2}(\Omega)}.
\end{aligned}$$

Noticing that $r_2 = \frac{N+p^-}{N} < p^*$. By the imbedding inequality and Young's inequality, we have

$$I_1 \le \varepsilon \int_\Omega |\nabla u|^{p^-} dx + c \le \varepsilon \int_\Omega \frac{|\nabla u|^{p(x)}}{p(x)} dx + c.$$

Now, we estimate $C_1 \int_\Omega (|u|^{\alpha+1} + |u_{i-1}|^{\alpha+1}) dx$ in the following two cases.

(i) $\alpha < p^- - 1$, which means that $\alpha < 1$. By Young inequality and Poincaré inequality, we get

$$C_1 \int_\Omega (|u|^{\alpha+1} + |u_{i-1}|^{\alpha+1}) dx \le \varepsilon \int_\Omega |u|^{p^-} dx + c \le \varepsilon \int_\Omega \frac{|\nabla u|^{p(x)}}{p(x)} dx + c.$$

(ii) $\alpha = p^- - 1 < 1$, and the Lebesgue measure of Ω is sufficiently small. By Poincaré inequality, we get

$$C_1 \int_\Omega |u|^{\alpha+1} + |u_{i-1}|^{\alpha+1} dx \le C_1 \int_\Omega |u|^{p^-} dx + c \le \varepsilon \int_\Omega \frac{|\nabla u|^{p(x)}}{p(x)} dx + c.$$

Summarizing up the above estimates and combining Lemma 1.1(vi), we get

$$\begin{aligned}
\psi^i(u) &\ge \frac{1}{2p^+} |\nabla u|^{p^-}_{L^{p(x)}(\Omega)} - C \\
&\ge \frac{1}{2Cp^+} |\nabla u|^{p^-}_{W^{1,p(x)}(\Omega)} - C \to \infty, \quad \text{as} \quad |u|_{W^{1,p(x)}(\Omega)} \to \infty.
\end{aligned}$$

3. $\psi(u)$ is weakly lower semicontinuous. At first, by the convexity of the functional, we know that for $\int_\Omega \frac{1}{p(x)}|\nabla u|^{p(x)}dx$, weakly lower semicontinuous is equivalent to lower semicontinuous (see [16]).

Let

$$v_l \to v, \text{ in } W_0^{1,p(x)} \text{ as } \quad l \to \infty. \tag{2.6}$$

Then by Lemma 1.1(iii), we have

$$|v_l|_{L^{p(x)}(\Omega)}, \int_\Omega |\nabla v_l|^{p(x)}dx \leq C, l = 1, 2, \cdots.$$

Now

$$\begin{aligned}
&| \int_\Omega \frac{1}{p(x)}|\nabla v_l|^{p(x)}dx - \int_\Omega \frac{1}{p(x)}|\nabla v|^{p(x)}dx \,| \\
&\leq \int_\Omega \int_0^1 |s\nabla v_l + (1-s)\nabla v|^{p(x)-1} \cdot |\nabla v_l - \nabla v| ds dx \\
&= \int_0^1 \int_\Omega |s\nabla v_l + (1-s)\nabla v|^{p(x)-1} \cdot |\nabla v_l - \nabla v| dx ds.
\end{aligned}$$

Then combining Lemma 1.1 (ii),(iii), we know that $\int_\Omega \frac{1}{p(x)}|\nabla u|^{p(x)}dx$ is a continuous functional.

Therefore, $\int_\Omega \frac{1}{p(x)}|\nabla u|^{p(x)}dx$ is weakly lower semicontinuous. Now, consider the functional

$$I_2 = -\int_\Omega F_i(x, u)dx + \frac{1}{2h}\int_\Omega (u - u_{i-1})^2 dx.$$

By (2.6), using (ii) and (iv) of Lemma 1.1, for any $0 < \varepsilon < p^-$, we have

$$v_l \rightharpoonup v, \quad \text{weakly in } W_0^{1,p^- - \epsilon},$$

and then using the Sobolev compact imbedding theorem we get

$$v_l \to v, \quad \text{in } L^{p_\epsilon^*},$$

where

$$p_\epsilon^* = \frac{N(p^- - \epsilon)}{N - (p^- - \epsilon)}.$$

For small enough ϵ, due to that $p^- > \frac{2N}{N+1}$, we have $p_\epsilon^* > \max\{\frac{r}{r-1}, 2\}$. Combining (A), we may prove that the functional I_2 is continuous in $L^{p_\epsilon^*}$. Hence I_2 is weakly lower semicontinuous.

Obviously, the sum of two weakly lower semicontinuous functionals is weakly lower semicontinuous functional and our conclusion follows. By above results and a standard argument (see [17]), we know that the functional $\psi^i(u)$ achieves its minimum on the set S. □

Let $u_+ = \max\{0, u\}$. Assume that u is a minima obtained in Lemma 2.1. Then for any constant $k \geq 1$, we have the following formulas (2.7)-(2.8) (see [1] p-7)

$$\begin{aligned}&\frac{1}{h}\int_\Omega (u-k)_+^2 dx + \int_{D_k} |\nabla u|^{p(x)} dx \\ &\leq \frac{1}{h}\int_\Omega (u-k)_+ \left(\int_{ih}^{(i+1)h} |f(x,\tau,u)|\, d\tau + |u_{i-1}|\right) dx,\end{aligned} \tag{2.7}$$

where

$$D_k = \{x \in \Omega : u(x) > k\}.$$

Moreover, if $u \in L^\infty(\Omega)$, we have

$$\begin{aligned}&\int_\Omega \frac{(u-k)_+^{q+1}}{h} dx + q\int_\Omega |\nabla u|^{p(x)} (u-k)_+^{q-1} dx \\ &\leq \int_\Omega \frac{(u_{i-1}-k)_+(u-k)_+^q}{h} dx + \frac{1}{h}\int_\Omega (u-k)_+^q \int_{ih}^{(i+1)h} f(x,\tau,u) d\tau dx,\end{aligned} \tag{2.8}$$

where $q \geq 1$ is a constant.

Now, we consider the following problem.

$$\frac{1}{h}(u_i - u_{i-1}) = \mathrm{div}(|\nabla u_i|^{p(x)-2}\nabla u_i) + \frac{1}{h}\int_{ih}^{(i+1)h} f(x,\tau,u_i) d\tau, x \in \Omega, \tag{2.9}$$

$$u_i \mid_{\partial\Omega} = 0,\ i = 1, 2, \cdots, \tag{2.10}$$

where $h > 0$ is a constant.

Lemma 7. *Let (A), (B) hold. Assume that* $p(x) \in C(\overline{\Omega})$ *and* $u_{i-1}(x) \in L^{p^*}(\Omega)$. *Then there exists a weak solution* u_i *of (2.9)-(2.10) such that* $u_i \in W_0^{1,p(x)}(\Omega)$.

Proof. By Lemma 2.1,for $0 < \epsilon < 1$ and $\eta \in C_0^\infty$,we have $u_i \pm \epsilon\eta \in S$, and then

$$g(-\epsilon) = \psi^i(u_i - \epsilon\eta) \geq \psi^i(u_i) = g(0),$$
$$g(\epsilon) = \psi^i(u_i + \epsilon\eta) \geq \psi^i(u_i) = g(0).$$

Therefore

$$\lim_{\epsilon\to 0^+} \frac{g(-\epsilon) - g(0)}{-\epsilon} \leq 0, \ \lim_{\epsilon\to 0^+} \frac{g(\epsilon) - g(0)}{\epsilon} \geq 0.$$

Plugging into the definition of g,we get

$$\int_\Omega \frac{1}{h}(u_i - u_{i-1})\eta dx$$
$$= -\int_\Omega \mathrm{div}\big(|\nabla u_i|^{p(x)-2}\nabla u_i\big)\cdot\nabla\eta dx + \int_\Omega \Big(\frac{1}{h}\int_{ih}^{(i+1)h} f(x,\tau,u_i)d\tau\Big)\eta dx, \ x \in \Omega.$$

Thus u_i is a weak solution of (2.9)-(2.10). □

3. THE PROOF OF MAIN RESULTS

Define $u^h : \Omega \times [0,\infty) \to R$ such that

$$u^{(h)}(\cdot,t) = u_i, \text{for} \quad t \in [ih,(i+1)h), i = 0,1,...,l, \tag{3.1}$$

where u_i is a solution obtained in Lemma 2.2, and l is an integer such that $lh \leq T < (l+1)h$.

We shall prove that a subsequence of $u^{(h)}$ converges and the limiting function is a solution of (1.1)-(1.3).

Denote

$$\partial^{(-h)}u^{(h)}(\cdot,t) = -\frac{1}{h}\left(u^{(h)}(\cdot,t-h) - u^{(h)}(\cdot,t)\right)$$
$$= \begin{cases} \frac{1}{h}(u_i - u_{i-1}), & \text{for } t \in [ih,(i+1)h), 1 \leq i \leq l, \\ 0, & \text{for } t \in [0,h). \end{cases} \tag{3.2}$$

Define the following new functions $f^{(h)}(x,t)$ and $\phi^{(h)}(x,t)$ as

$$f^{(h)}(x,t) = \frac{1}{h}\int_{ih}^{(i+1)h} f(x,\tau,u_i(x))d\tau, \quad \text{for} t \in [ih,(i+1)h), i = 0,1,...,l, \tag{3.3}$$

$$\phi^{(h)}(x,t) = \frac{1}{h}\int_{ih}^{(i+1)h} \phi(x,\tau)d\tau, \quad \text{for} \quad t \in [ih, (i+1)h), i = 0, 1, ..., l, \tag{3.4}$$

By (A) we have

$$|f^{(h)}(x,t)| \leq C_0(\phi^{(h)} + |u^{(h)}|^{\alpha}). \tag{A'}$$

Remark 8. *(see [1] p-8) If $\phi \in L^r(Q_T)$, then $\phi^{(h)} \in L^r(Q_T)$ and*

$$\iint_{Q_T} (\phi^{(h)})^r dxdt \leq \iint_{Q_T} \phi^r dxdt,$$

where r is given in (A).

In the following, we shall give the estimate to the maximum norm of the solution by adopting the method in [1].

Lemma 9. *Let (A), (B) hold. Assume that*

$$p(x) \in C(\overline{\Omega}) \ \textit{and} \ u_0 \in L^{\infty}(\Omega) \cap W_0^{1,p(x)}(\Omega),$$

Then for any integer $1 \leq q < \infty$, there is a constant $C(q) > 0$ independent of h such that

$$\|u^{(h)}\|_{L^{q+1}(Q_T)} \leq C(q), \quad \forall h > 0.$$

Proof. Let $u_+ = \max\{0, u\}$ and k be chosen so that $\|u_0\|_{L^\infty(\Omega)} \leq k$. Multiplying (3.1) by $(q+1)(u_i - k)_+^q$ and integrating over Ω we get

$$\begin{aligned}
&(q+1)\int_\Omega \frac{(u_i-k)_+^{q+1}}{h}dx + q(q+1)\int_\Omega |\nabla u_i|^{p(x)}(u_i-k)_+^{q-1}dx \\
&= (q+1)\int_\Omega \frac{(u_{i-1}-k)(u_i-k)_+^q}{h}dx + (q+1)\int_\Omega (u_i-k)_+^q f^{(h)}(x,ih)dx \\
&\leq (q+1)\int_\Omega \frac{(u_{i-1}-k)_+(u_i-k)_+^q}{h}dx + (q+1)\int_\Omega (u_i-k)_+^q f^{(h)}(x,ih)dx
\end{aligned} \tag{3.5}$$

By Young inequality

$$\begin{aligned}
&(u_{i-1}-k)_+(u_i-k)_+^q \\
&\leq \frac{q}{q+1}(u_i-k)_+^{q+1} + \frac{1}{q+1}(u_{i-1}-k)_+^{q+1}.
\end{aligned}$$

Hence, we have

$$\begin{aligned}&\int_\Omega \frac{(u_i-k)_+^{q+1}}{h}dx+q(q+1)\int_\Omega |\nabla u_i|^{p(x)}(u_i-k)_+^{q-1}dx\\&\le \int_\Omega \frac{(u_i-k)_+^{q+1}}{h}dx+(q+1)\int_\Omega (u_i-k)_+^q f^{(h)}(x,ih)dx, i=1,2,...,l.\end{aligned} \tag{3.6}$$

Summing over i in (3.6) and considering the definition of $u^{(h)}$, we have

$$\begin{aligned}&\int_\Omega (u^{(h)}-k)_+^{q+1}(\cdot,t)dx+q(q+1)\int_h^{(l+1)h}\int_\Omega |\nabla u^{(h)}|^{p(x)}(u^{(h)}-k)_+^{q-1}dxdt\\&\le \int_\Omega (u_0-k)_+^{(q+1)}dx+(q+1)\int_h^{(l+1)h}\int_\Omega (u^h-k)_+^q f^{(h)}dxdt,\end{aligned} \tag{3.7}$$

where $t\in[h,(l+1)h)$. By Young inequality

$$|\nabla u^{(h)}|^{p^-}\le |\nabla u^{(h)}|^{p(x)}+C.$$

Using $\|u_0\|_{L^\infty(\Omega)}\le k$, we get

$$\begin{aligned}&\sup_{t\in(h,(l+1)h)}\int_\Omega (u^{(h)}-k)_+^{(q+1)}(\cdot,t)dx+q(q+1)\int_h^{(l+1)h}\int_\Omega |\nabla u^h|^{p^-}(u^h-k)_+^{q-1}dxdt\\&\le Cq(q+1)\int_h^{(l+1)h}\int_\Omega (u^{(h)}-k)_+^{q-1}dxdt+(q+1)\int_h^{(l+1)h}\int_\Omega (u^{(h)}-k)_+^q|f^{(h)}|dxdt\\&=Cq(q+1)I_1+(q+1)I_2.\end{aligned} \tag{3.8}$$

Denote

$$\mu(k)=|\{(x,t)\in\Omega\times(0,(l+1)h):u^{(h)}\ge k\}|.$$

By Young inequality and the Poincaré inequality, we get

$$\begin{aligned}I_1&\le \int_h^{(l+1)h}\int_\Omega (u^{(h)}-k)_+^{q+p^--1}dxdt+C(q)\mu(k)\\&\le C(|\Omega|)\int_h^{(l+1)h}\int_\Omega |\nabla (u^{(h)}-k)_+^{\frac{q+p^--1}{p^-}}|^{p^-}dxdt+C(q)\mu(k).\end{aligned} \tag{3.9}$$

Now we estimate I_2,

$$I_2 \leq C_0 \int_h^{(l+1)h} \int_\Omega (u^{(h)} - k)_+^q (\phi^{(h)} + |u^{(h)} - k + k|^\alpha) dx d\tau.$$

Since $\alpha < 1$, for any $c > 1$, we have

$$|u^{(h)} - k + k|^\alpha \leq c(|u^{(h)} - k|^\alpha + ck^\alpha),$$

then,

$$I_2 \leq C[\int_h^{(l+1)h} \int_\Omega (u^{(h)} - k)_+^q \phi^{(h)} dx d\tau + \int_h^{(l+1)h} \int_\Omega |u^{(h)} - k|^{q+\alpha} dx d\tau].$$

By Hölder inequality and Poincaré inequality

$$\begin{aligned} &\int_h^{(l+1)h} \int_\Omega (u^{(h)} - k)_+^{q+\alpha} dx dt \\ &\leq C \int_h^{(l+1)h} (\int_\Omega (u^{(h)} - k)_+^{q+p^- -1} dx)^{\frac{q+\alpha}{q+p^- -1}} dt \qquad (3.10) \\ &\leq C \int_h^{(l+1)h} (|\nabla (u^{(h)})_+^{\frac{q+p^- -1}{p^-}}|^{p^-} dx)^{\frac{q+\alpha}{q+p^- -1}} dt. \end{aligned}$$

At the same time, similar as the proof of (3.10), using the imbedding theorem, we may prove (see Lemma 2.1 in [1]) that

$$\begin{aligned} &\int_h^{(l+1)h} \int_\Omega (u^{(h)} - k)_+^q \phi^{(h)} dx dt \\ &\leq C(sup_{t\in(h,(l+1)h)} \int_\Omega (u^{(h)} - k)_+^{(q+1)}(\cdot, t) dx \qquad (3.11) \\ &+ \int_h^{(l+1)h} \int_\Omega |\nabla (u^{(h)} - k)_+^{\frac{(q+p^- -1)}{p^-}}|^{p^-} dx dt)^{q_1}, \end{aligned}$$

where $q_1 = \frac{q(N+p^-)}{(q(N+p^-)+N(p^- -1+p^-/N))} < 1$.

Combining (3.10), (3.11), Remark 3.1 and (A'), we can obtain the estimate for I_2.

Substituting it into (3.8), by $|\mu(k)| \leq 2|Q_T|$ and Young inequality, we get

$$\sup_{t\in(0,(l+1)h)} \int_\Omega (u^{(h)} - k)_+^{(q+1)}(\cdot, t) dx \leq C. \qquad (3.12)$$

Here we used the fact that $(u^{(h)} - k)_+(\cdot, t) = 0$, for $t \in [0, h)$.

If $\alpha = p^- - 1$ and $|\Omega|$ is sufficiently small, by Poincaré inequality,

$C(|\Omega|) \to 0$, as $|\Omega| \to 0$, in (3.9) and (3.10).

Thus we can also obtain the estimate for I_2. Substituting it into (3.8), we can prove (3.12).

Similarly, we may prove

$$\sup_{t\in(0,(l+1)h)} \int_\Omega (-u^{(h)} - k)_+^{(q+1)}(\cdot, t)dx \le C.$$

Thus

$$||u^{(h)}||_{L^{q+1}(Q_T)} \le C(q).$$

□

Remark 10. *If we take $k = 0$ and $q = 1$ in Lemma 3.2, then*

$$\int_h^{(l+1)h} \int_\Omega |\nabla u^{(h)}|^{p(x)} dxdt \le C. \tag{3.13}$$

Remark 11. *For studying above problems, we have to study the terms like $\int |\nabla u_i|^{p(x)}|u_i|^q dx (q \ge 0)$ for $|\nabla u_i| \in L^{p(x)}$(see (3.5)). To insure that integrals to be well-defined, we need $u_i \in L^\infty$. Actually, the solution that we get in Lemma 2.1 can be considered as a bounded function (see Lemma 3.7).*

Now, we give a uniform estimate to the maximum norm of the solution. We need the following proposition.

Remark 12. *(See [4] p-12.)Let $Y_n, n = 0, 1, 2, \cdots$, be a sequence of positive numbers,satisfying the recursive inequalities*

$$Y_{n+1} \le Bb^n Y_n^{1+\beta}$$

where $B, b > 1$ and $\beta > 0$ are given numbers. If

$$Y_0 \le B^{-1/\beta} b^{-1/\beta^2}, \tag{3.14}$$

then $\{Y_n\}$ converges to zero as $n \to \infty$.

Lemma 13. *Let the assumptions of Lemma 3.2 hold. Then there is a constant $M_1 > 0$ depending only on $T, |\Omega|, N, p^-, r, ||u_0||_{L^\infty(Q_T)}$ such that*

$$||u^{(h)}||_{L^\infty(Q_T)} \leq M_1, \forall h > 0.$$

Proof. Let k be chosen so that $||u_0||_{L^\infty(\Omega)} \leq k$ and denote

$$J_k = \sup_{t\in(0,(l+1)h)} \int_\Omega (u^{(h)} - k)_+^2(\cdot, t)dx + \int_0^{(l+1)h} \int_\Omega |\nabla(u^{(h)} - k)_+|^{p^-} dxdt.$$

Take $q = 1$ in (3.8), then by (A')

$$J_k \leq C_1(\int_0^{(l+1)h} \int_\Omega (\phi^{(h)} + |u^{(h)}|^\alpha)(u^{(h)} - k)_+ dxdt + \mu(k)). \quad (3.15)$$

Now we estimate the integral of the right-hand side of the inequality.

By Remark 3.1 and Hölder's inequality, we get

$$\begin{aligned}
&\int_0^{(l+1)h} \int_\Omega \phi^{(h)}(u^{(h)} - k)_+ dxdt \\
&\leq C_2\left(\int_0^{(l+1)h} \int_\Omega (u^{(h)} - k)_+^{\frac{r}{r-1}} dxdt\right)^{\frac{r-1}{r}} \\
&\leq C_2\left(\int_0^{(l+1)h} \int_\Omega (u^{(h)} - k)_+^{p^- + \frac{2p^-}{N}} dxdt\right)^{\frac{N}{p^- N+2p^-}} \mu(k)^{\frac{r}{r-1} - \frac{N}{p^- N+2p^-}}.
\end{aligned}$$

Hence, by the imbedding inequality (see [4] p-7 or [17] p-62) we have

$$\begin{aligned}
&\left(\int_0^{(l+1)h} \int_\Omega (u^{(h)} - k)_+^{p^- + \frac{2p^-}{N}} dxdt\right)^{\frac{N}{p^- N+2p^-}} \\
&\leq \left(\int_0^{(l+1)h} \int_\Omega \nabla(u^{(h)} - k)_+^{p^-} dxdt\right)^{\frac{N}{p^- N+2p^-}} \\
&\cdot \left(\text{ess sup}_{0<t<T} \int_\Omega (u^{(h)} - k)_+^2 dx\right)^{\frac{p}{p^- N+2p^-}}.
\end{aligned}$$

By Young inequality, we have,

$$\int_0^{(l+1)h} \int_\Omega (u^{(h)} - k)_+ \phi^{(h)} dxdt \leq C_3 J_k^{\frac{N+p^-}{p^- N+2p^-}} \mu(k)^{\frac{r}{r-1} - \frac{N}{p^- N+2p^-}}. \quad (3.16)$$

Also, by Lemma 3.2 we have

$$\begin{aligned}
&\int_0^{(l+1)h}\int_\Omega |u^{(h)}|^\alpha (u^{(h)}-k)_+ dxdt \\
&\le \left(\int_0^{(l+1)h}\int_\Omega \left(u^{(h)}\right)^{\alpha r} dxdt\right)^{\frac{1}{r}} \left(\int_0^{(l+1)h}\int_\Omega (u^{(h)}-k)_+^{\frac{r}{r-1}} dxdt\right)^{\frac{r-1}{r}} \\
&\le C_5\left(\int_0^{(l+1)h}\int_\Omega (u^{(h)}-k)_+^{\frac{r}{r-1}} dxdt\right)^{\frac{r-1}{r}} \\
&\le C_6 J_k^{\frac{N+p^-}{p^-N+2p^-}} \mu(k)^{\frac{r}{r-1}-\frac{N}{p^-N+2p^-}}.
\end{aligned} \tag{3.17}$$

Substituting (3.16), (3.17) into (3.15), we get

$$J_k \le C_7 J_k^{\frac{N+p^-}{p^-N+2p^-}} \mu(k)^{\frac{r}{r-1}-\frac{N}{p^-N+2p^-}} + C_7\mu(k).$$

By Young inequality

$$J_k \le C_8\left(\mu(k)^{1+\frac{(r-N-2)p^-}{(rN(p^--1)+rp^-)}} + \mu(k)\right).$$

Hence, for all $k_2 \ge k_1$, we have

$$\begin{aligned}
&(k_2-k_1)\mu(k_2) \\
&= \int_0^{(l+1)h}\int_{D_{(k_2)}} (k_2-k_1)dxdt \\
&\le \int_0^{(l+1)h}\int_{D_{(k_2)}} (\mu_h-k_1)_+ dxdt \\
&\le C_9\gamma^{\frac{N}{p^-N+2p^-}} J_{k_1}^{\frac{N+p^-}{p^-N+2p^-}} \\
&\le C_{10}\left(\mu(k_1)^{1+\frac{(r-N-2)p^-}{rN(p^--1)+rp^-}} + \mu(k_1)\right)^{\frac{N+p^-}{p^-N+2p^-}}.
\end{aligned} \tag{3.18}$$

where γ is a constant, depending only on N, p^-, T, comes from the imbedding inequality (see [4] p-7 or [17] p-62).

If we take

$$k_2 = \|u_0\|_{L^\infty(\Omega)} + j (j > 1)$$

and

$$k_1 = \|u_0\|_{L^\infty(\Omega)} + 1,$$

then

$$\mu(k_2)^{\frac{N}{p^- N + 2p^-}} \leq \frac{C_{10}}{j-1}\left(((T+1)|\Omega|)^{1+\frac{(r-N-2)p^-}{rN(p^- -1)+rp^-}} + (T+1)|\Omega| \right)^{\frac{N+p^-}{p^- N+2p^-}}$$

Hence, there exists a constant $j_0 > 1$ depending only on T , $|\Omega|$, N, p^-, r such that

$$\mu(k_2) \leq 1, as \quad j \geq j_0.$$

We take

$$k_m = \tilde{M}(2 - 2^{-m}), m = 0, 1, 2, \cdots,$$

where $\tilde{M} \geq \|u_0\|_{L^\infty(\Omega)} + j_0$ is a constant. Then it is easy to see that

$$\mu(k_m) \leq 1, for \quad m = 0, 1, 2, \cdots. \tag{3.19}$$

Now we consider the following two cases.

If $(i)\, 1 + \frac{(r-N-2)p^-}{rN(p^- -1)+rp^-} \geq 1$,
then by (3.18) and (3.19)

$$(k_{m+1} - k_m)\big(\mu(k_{m+1})\big)^{\frac{N}{p^- N+2p^-}} \leq C_{11}\big(\mu(k_m)\big)^{\frac{N+p^-}{p^- N+2p^-}},$$

i.e.

$$\mu(k_{m+1}) \leq \left(\frac{2C_{11}}{\tilde{M}}\right)^{\frac{(p^- N+2p^-)}{N}} 2^{\frac{m(p^- N+2p^-)}{N}} \mu(k_m)^{1+\frac{p^-}{N}}.$$

If $(ii)\, 1 + \frac{(r-N-2)p^-}{rN(p^- -1)+rp^-} < 1$, then

$$(k_{m+1} - k_m)(\mu(k_{m+1}))^{\frac{N}{p^- N+2p^-}} \leq C_{12}\left(\mu(k_m)^{\frac{1+(r-N-2)p^-}{rN(p^- -1)+rp^-}} \right)^{\frac{N+p^-}{p^- N+2p^-}},$$

i.e.

$$\mu(k_{m+1}) \leq \left(\frac{2C_{12}}{\tilde{M}}\right)^{\frac{(p^- N+2p^-)}{N}} 2^{\frac{m(p^- N+2p^-)}{N}} \mu(k_m)^{1+\delta_1},$$

where

$$\delta_1 = \frac{p^-}{N} + \frac{(r-N-2)p^-(N+p^-)}{(rN(p^- - 1) + rp^-)N} > 0.$$

Now, taking

$$\tilde{M} = max\{C_{11}2^{\frac{N}{p^- + 1}}, C_{12}2^{\frac{N}{p^- - \delta_1(N+2)} + 1}, \|u_0 \|_{L^\infty(\Omega)} + j_0\},$$

then in both cases (3.14) always holds. Hence, by Remark 3.5, we have

$$u^{(h)} \leq 2\tilde{M}.$$

Similarly, we may derive a lower bound and this completes the proof of Lemma 3.6. □

Up to now we require that $p(x) \in C(\overline{\Omega})$. This is in fact not necessary. We have

Lemma 14. *Assume that* $u_0 \in L^\infty(\Omega) \cap W_0^{1,p(x)}(\Omega)$ *and (A), (B) hold. Then the conclusions in Lemma 2.1, Remark 3.1, Lemma 3.2, Remark 3.3, Lemma 3.6 still hold.*

Proof. Note that we need the condition $p(x) \in C(\overline{\Omega})$ only in Lemma 1.1. But, if the functions mentioned in the proofs are uniformly bounded, then Lemma 1.1 will still holds without continuity condition.

Now, replace S in (2.4) by

$$\tilde{S} = \{u \in W_0^{1,p(x)}(\Omega)\} \cap \{\|u\|_{L^\infty(\Omega)} \leq M_1 + 1\}, \tag{3.20}$$

and still consider $\psi^i(u)$, where $M_1 > \|u_0\|_{L^\infty(\Omega)}$ is a constant depending only on $T, |\Omega|, N, p^-, r, N, \|u_0\|_{L^\infty(Q_T)}$ (the same as that in Lemma 3.6).

Assume that

$$v_l \in \tilde{S} \quad and \quad v_l \xrightarrow{weak} v \quad in \quad W_0^{1,p(x)}(\Omega),$$

then by (ii) and (iv) of Lemma 1.1, we have $v_l \xrightarrow{weak} v$ in $W_0^{1,p(x)}(\Omega)$. Hence, $\|v\|_{L^\infty(\Omega)} \leq M_1 + 1$ and $\tilde{S}$ is weakly closed.

So we can complete the proof of Lemma 1.1 without the condition $p(x) \in C(\overline{\Omega})$. On the other hand, the proof of Remark 1.4 does not need any change. Note that by the definition of $\tilde{S}$, it is not clear whether the minima of $\psi^i(u)$

satisfies (2.9) or not. Hence, we need modify the proofs of Lemmas 3.2 and 3.6 Let u_i be a minima of $\psi^i(u)$ in $\tilde{S}$. Then for

$$0 \leq \epsilon < (1/(\|u_i\|_{L^\infty(\Omega)} + 1)^{q-1}),$$

we have

$$\|u_i - \varepsilon(u_i - k)_+^q\|_{L^\infty(\Omega)} \leq M_1 + 1,$$

and

$$u_i - \varepsilon(u_i - k)+^q \in \tilde{S}.$$

This estimate is exactly the same as that (3.5) in Lemma 3.2. Then without any change in Lemmas 3.2 and 3.6, we get

$$\|u_i\|_{L^\infty(\Omega)} \leq M_1.$$

Thus for any $\phi \in C_0^\infty(\Omega)$, $|\phi| \leq 1$ and $-1 \leq \varepsilon \leq 1$, we have that $u_i + \varepsilon\phi \in \tilde{S}$ and then

$$\psi^i(u_i + \varepsilon\phi) \geq \psi^i(u_i).$$

We can prove that u_i is a weak solution of (2.9)-(2.10). We must point out that the test function must satisfy $|\phi| \leq 1$. Considering the form the test function appeared in the equality, the constrain to ϕ may be removed. □

Lemma 15. *Let the assumptions of Lemma 3.7 hold. Then for any integer $1 \leq \tilde{l} \leq l$, we have*

$$\begin{aligned} &\frac{1}{2}\int_0^{(\tilde{l}+1)h}\int_\Omega \left|\partial^{(-h)}u^{(h)}\right|^2 dxdt + \int_\Omega \frac{1}{p(x)}\left|\nabla u^{(h)}(x,\tilde{l}h)\right|^{p(x)} dx \\ &\leq \int_\Omega \frac{1}{p(x)}|\nabla u_0|^{p(x)} dx. \end{aligned} \tag{3.21}$$

Proof. From Lemma 3.7, we know that u_i is the minima of $\psi^i(u)$. Hence for

$$u_{i-1} \in \tilde{S}, \psi^i(u_i) \leq \psi^i(u_{i-1}),$$

and then

$$\begin{aligned} &\int_\Omega \frac{1}{p(x)}|\nabla u_i|^{p(x)} dx + \int_\Omega \frac{1}{2h}|u_i - u_{i-1}|^2 dx \\ &\leq \int_\Omega \frac{1}{p(x)}|\nabla u_{i-1}|^{p(x)} dx + \int_\Omega \int_{u_{i-1}}^{u_i} (\frac{1}{h}\int_{ih}^{(i+1)h} f(x,\tau,s)d\tau)dsdx, \end{aligned}$$

where $i = 1, 2, \cdots, \tilde{l}$.

Summing over i, we have

$$\int_\Omega \frac{1}{p(x)} |\nabla u_{\tilde{l}}|^{p(x)} dx + \sum_{i=1}^{\tilde{l}} \frac{1}{2h} \int_\Omega |u_i - u_{i-1}|^2 dx$$
$$\leq \int_\Omega \frac{1}{p(x)} |\nabla u_0|^{p(x)} dx + \sum_{i=1}^{\tilde{l}} \int_\Omega \int_{u_{i-1}}^{u_i} (\frac{1}{h} \int_{ih}^{(i+1)h} f(x,\tau,s) d\tau) ds dx. \tag{3.22}$$

We estimate the second term in the inequality in what follows. By Lemma 3.2, Young inequality and the differentiability of f , we get

$$\int_\Omega \int_{u_{i-1}}^{u_i} (\frac{1}{h} \int_{ih}^{(i+1)h} f(x,\tau,s) d\tau) ds dx$$
$$\leq C \int_\Omega |u_i - u_{i-1}| dx \leq \frac{1}{3h} \int_\Omega |u_i - u_{i-1}|^2 dx + 3Ch.$$

Plugging into (3.22), we get

$$\begin{aligned} &\int_\Omega \frac{1}{p(x)} |\nabla u_l|^{p(x)} dx + \sum_{i=1}^{\tilde{l}} \frac{1}{3h} \int_\Omega |u_i - u_{i-1}|^2 dx \\ &\leq \int_\Omega \frac{1}{p(x)} |\nabla u_0|^{p(x)} dx + 3CT. \end{aligned} \tag{3.23}$$

The conclusion of the lemma follows by noticing the definition of $u^{(h)}$. □

Define a new function:

$$w^{(h)}(\cdot,t) = \begin{cases} (\frac{t}{h} - i)(u_i - u_{i-1}) + u_{i-1}, t \in [ih, (i+1)h), i = 1, 2, ..., l, \\ u_0, \qquad t \in [0,h). \end{cases}$$

Then, we have

Lemma 16. *Let the assumptions of Lemma 3.7 hold. Then*

$$\int_0^T \int_\Omega \left| w^{(h)} - u^{(h)} \right|^2 dx dt \to 0, \quad as \quad h \to 0.$$

Proof. By a direct calculation,

$$\begin{aligned}
&\int_\Omega \int_{ih}^{(i+1)h} \left| w^{(h)} - u^{(h)} \right|^2 dxdt \\
&= \int_\Omega (u_i - u_{i-1})^2 dx \int_{ih}^{(i+1)h} (\frac{t}{h} - i - 1)^2 dt \\
&= \frac{h}{2} \int_\Omega (u_i - u_{i-1})^2 dx \\
&= \frac{h^2}{2} \int_\Omega \int_{ih}^{(i+1)h} (\partial^{-h} u^{(h)})^2 dxdt.
\end{aligned}$$

Summing over i and using Lemma 3.8,we get

$$\int_h^T \int_\Omega \left| w^{(h)} - u^{(h)} \right|^2 dxdt \to 0, \quad as \quad h \to 0.$$

On the other hand,

$$\int_0^T \int_\Omega \left| w^{(h)} - u^{(h)} \right|^2 dxdt = 0.$$

The conclusion follows. □

Lemma 17. *Let the assumptions of Lemma 3.7 hold. Then there exists a subsequence of $\{u^{(h)}\}$ denoted again by itself for the sake of simplicity, and a function u such that*

$$u^{(h)} \to u, \quad in \quad L^{\frac{2N}{N+1}}(Q_T), \tag{3.24}$$

$$u^{(h)} \to u, \quad a.e. \quad in \quad Q_T, \tag{3.25}$$

$$\partial^{-h} u^{(h)} \underrightarrow{weak}\ \partial_t u, \quad in \quad L^2(Q_T), \tag{3.26}$$

$$\nabla u^{(h)} \underrightarrow{weak}\ \nabla u, \quad in \quad L^{p(x)}(Q_T), \tag{3.27}$$

as $h \to 0$.

Proof. Let $r = \frac{2N}{N+1}$. By Lemma 3.7 and Young inequality,

$$\begin{aligned}
&\int_\Omega \left| \nabla u^{(h)}(\cdot, t) \right|^r dx \\
&\le \int_\Omega \left(\frac{r |\nabla u^{(h)}(\cdot, t)|^{p(x)}}{p(x)} + \frac{p(x) - r}{p(x)} \right) dx \qquad (3.28) \\
&\le C + C \int_\Omega \frac{1}{p(x)} \left| \nabla u^{(h)}(\cdot, t) \right|^{p(x)} dx \le C.
\end{aligned}$$

By Poincaré inequality,

$$\int_\Omega \left|u^{(h)}(\cdot,t)\right|^{\frac{2N}{N+1}} dx \le C,$$

and then

$$\int_0^T \int_\Omega \left|u^{(h)}(\cdot,t)\right|^{\frac{2N}{N+1}} dxdt \le CT. \tag{3.29}$$

Therefore, there exists a subsequence of $u^{(h)}$ (denoted again by itself) and a function u such that

$$u^{(h)} \underrightarrow{weak} \quad u, \quad in \quad L^{\frac{2N}{N+1}}(Q_T). \tag{3.30}$$

And then there exists a subsequence of $u^{(h)}$ such that (3.25) holds.

In the following, we consider $w^{(h)}$. Since

$$\nabla w^{(h)} = (\nabla u_i - \nabla u_{i-1})(\frac{t}{h} - i) + \nabla u_{i-1},$$

$$t \in [ih, (i+1)h), i = 1, 2, \cdots, l,$$

by (3.28), (3.29) and Lemma 3.9, we know that $w^{(h)}$ and $\nabla w^{(h)}$ are uniformly bounded in $L^{\frac{2N}{N+1}}(Q_T)$. Since

$$w_t^{(h)} = \partial^{(-h)} u^{(h)} = \begin{cases} \frac{1}{h}(u_i - u_{i-1}), for & t \in [ih, (i+1)h), i = 1, 2, ..., l,, \\ 0, \qquad\qquad for & t \in [0, h), \end{cases}$$

by Lemma 3.8, we have $w_t^{(h)} \in L^2(Q_T)$. By the above estimates, we know that there exists a subsequence of $w^{(h)}$ (denoted again by itself) and a function u_* such that

$$w^{(h)} \to u_*, \quad in \quad L^{\frac{2N}{N+1}}(Q_T),$$

$$\nabla w^{(h)} \underrightarrow{weak} \nabla u_*, \quad in \quad L^{\frac{2N}{N+1}}(Q_T),$$

$$\partial^{(-h)} u^{(h)} = w_t^{(h)} \underrightarrow{weak} (u_*)_t, \quad in \quad L^2(Q_T).$$

By Lemma 3.6, we get $u = u_*$. Using this, it is easy to prove (3.24) and (3.27). By Lemma 1.1, $(L^{p(x)}, |\cdot|_{p(x)})$ is weakly compact and hence by Lemma 3.8, (3.27) can be easily obtained. □

Remark 18. *By (3.21), we know that* $u \in L^{\infty}(0, T; W_0^{1,p(x)}(\Omega))$.

Lemma 19. *Let the assumptions of Lemma 3.7 hold. Then*

$$f^{(h)} \to f(x, t, u),\ in\ L^1(Q_T),\ as\, h \to 0. \tag{3.31}$$

Proof. By Remark 3.3 and Lemma 3.7, we have

$$\|u^{(h)}\|_{L^{\infty}(Q_T)}, \|u\|_{L^{\infty}(Q_T)} \leq C, \quad \forall h > 0.$$

Since $f \in C^1$, by Hölder's inequality,

$$\begin{aligned}
&\iint_{Q_T} \left| f^{(h)} - f \right| dxdt \\
&\leq \sum_i \int_{ih}^{(i+1)h} \int_{\Omega} \frac{1}{h} \int_{ih}^{(i+1)h} \left| f(x, \tau, u_i) - f(x, t, u) \right| d\tau dxdt \\
&\leq C \sum_i \int_{ih}^{(i+1)h} \int_{\Omega} \frac{1}{h} \int_{ih}^{(i+1)h} ||\tau - t| + |u_i - u|| d\tau dxdt \\
&\leq C \sum_i \int_{ih}^{(i+1)h} \int_{\Omega} \frac{1}{h} \int_{ih}^{(i+1)h} h d\tau dxdt + C \sum_i \int_{ih}^{(i+1)h} \int_{\Omega} |u_i - u|\, dxdt \\
&\leq C \int_0^T \int_{\Omega} h dxdt + C \int_0^T \int_{\Omega} \left| u^{(h)} - u \right| dxdt \to 0, \quad as \quad h \to 0.
\end{aligned}$$

□

Proof of Theorem 1.1. By Lemma 3.8 and the weak compactness of the space there exists a subsequence such that

$$\left| \nabla u^{(h)} \right|^{p(x)-2} u_{x_i}^{(h)} \underrightarrow{weak}\ \chi_i, \quad in\ L^{\frac{p(x)}{p(x)-1}}(Q_T).$$

The same as that in [1], we may prove that

$$\chi_i = |\nabla u|^{p(x)-2} u_{x_i}.$$

For test function $\phi(x, t) \in C_0^{\infty}(Q_T)$ and any constant $\tilde{\tau} \in [0, T]$, we have $\phi(x, \tilde{\tau}) \in C_0^{\infty}(\Omega)$. Hence, by Lemma 2.2,

$$\int_{\Omega} \partial^{(-h)} u^{(h)} \phi(x, \tilde{\tau}) dx = \int_{\Omega} |\nabla u_i|^{p(x)-2} \nabla u_i \nabla \phi(x, \tilde{\tau}) dx + \int_{\Omega} f^{(h)} \phi(x, \tilde{\tau}) dx$$

Integrating for $\tilde{\tau}$, combining Lemmas 3.10, Lemma 3.12 and Remark 3.3, we may prove that u is a weak solution of equation (1.1).

Now we prove that u satisfies the initial condition, i.e.(1.6) holds.

In the problem (2.9)-(2.10), taking a test function $\tilde{\psi}(x) \in C_0^\infty(\Omega)$, we get

$$\begin{aligned} &\int_\Omega (u_i - u_{i-1})\tilde{\psi}dx + \int_{ih}^{(i+1)h} dt \int_\Omega |\nabla u_i|^{p(x)-2}\nabla u_i \cdot \nabla\tilde{\psi}dx \\ &= \int_\Omega \left(\int_{ih}^{(i+1)h} f(x,\tau,u_i)d\tau \right) \tilde{\psi}dx, \quad i = 1, 2, \cdots . \end{aligned} \tag{3.32}$$

Summing over i, we get

$$\begin{aligned} &\int_\Omega (u_{\tilde{l}} - u_0)\tilde{\psi}dx \\ &= -\int_h^{(\tilde{l}+1)h} \int_\Omega |\nabla u_i|^{p(x)-2}\nabla u_i \cdot \nabla\tilde{\psi}dxdt + \sum_i h \int_\Omega (\int_{ih}^{(i+1)h} f(x,\tau,u_i)d\tau)\tilde{\psi}dx, \end{aligned}$$

where $\tilde{l} > 0$ is an integer. Then by (3.13) and (ii), (iii) of Lemma 1.1, we have

$$\left| \int_h^{(\tilde{l}+1)h} \int_\Omega |\nabla u_i|^{p(x)-2}\nabla u_i \cdot \nabla\tilde{\psi}dxdt \right| \le C(\sup\left|\nabla\tilde{\psi}\right|)|\nabla u_i\ _{p(x)}|1|_{p(x)} \le C(\tilde{l}h)^{\delta_1},$$

where $\delta_1 > 0$ is a constant depending only on p^+ and p^-.

By (A), combining Poincaré inequality and Hölder's inequality

$$\left| \int_\Omega (u_{\tilde{l}} - u_0)\tilde{\psi}dx \right| \le C(\tilde{l}h)^{\delta_1} + \tilde{l}h + (\tilde{l}h)^{\frac{N}{N+p^-}} + (\tilde{l}h)^{\frac{1-\alpha}{p^-}}.$$

For $\tilde{l}h < 1$, there exists a constant $\delta_2 > 0$ depending only on α, p^-, N and δ_1 such that

$$\left| \int_\Omega (u_{\tilde{l}} - u_0)\tilde{\psi}dx \right| \le C(\tilde{l}h)^{\delta_2}.$$

Noticing the definition of the $u^{(h)}$, we have

$$\sup_{t\in[h,(\tilde{l}+1)h)} \left| \int_\Omega (u^{(h)}(x,t) - u_0)\tilde{\psi}dx \right| \le Ct^{\delta_2}. \tag{3.33}$$

For $t \in [0, h)$,

$$\int_\Omega (u^{(h)}(x,t) - u_0)\tilde{\psi}dx = 0.$$

Hence for all $0 \leq t < 1$, (3.33) holds. By (3.25), letting $h \to 0$ in (3.33), we may easily get (1.6).

At the last of the paper, we are able to prove that, similar as that in [1], the uniqueness of the solutions is true. We omit the details here.

ACKNOWLEDGMENTS

The authors wish to thank the referee for his (her) careful reading and valuable remarks which improve the presentation of the chapter. The author was supported in part by Fujian Education and Scientific Research Project for Young and Middle-aged Teachers JT180773.

CONCLUSION

In this chapter we consider a $p(x)$-Laplacian equation with the Dirichlet boundary condition. The existence and uniqueness of weak solutions are obtained provided that $ess\,\inf p(x) > \frac{2N}{N+1}$,which improves the results of Lian etc. [1]. The result obtained in this chapter might find some potential applications in nonlinear fluid dynamics.

REFERENCES

[1] Lian S., Gao W., Yuan H., Cao Ch., Existence of solutions to an initial Dirichlet problem of evolutional p(x)-Laplace equations, *Ann. I. H. Poincare-AN*, (29)2012, pp. 377-399.

[2] Oleinik O. A., *Linear equations of second order with nonnegative characteristic form*, Math. sb, 609(1966), pp. 111-140; English transl.: Amer. Math. Soc. Tranl, 65(1967), pp.167-199.

[3] Zhao J. N., *Existence and nonexistence of solutions for* $u_t = div(|\nabla u|^{p-2}\nabla u) + f(\nabla u, u, x, t)$, J. Math. Anal. Appl, 172(1993), pp. 130-146.

[4] Dibenedetto E., *Degenerate Parabolic Equations*, Springer-Verlag, New York, 1993.

[5] Zhao J. N., On the Cauchy problem and initial traces for the evolution p-laplacian equations with strongly nonlinear sources, *J. Diff. Eqs.* 121(1995), pp. 329-383.

[6] Ruzicka M., Electrorheological Fluids: Modeling and Mathematical Theory, *Lecture Notes in Math.*, vol. 1748, Springer, Berlin, 2000.

[7] Acerbi E., Mingione G., Regularity results for stationary electrorheological fluids, *Arch. Ration. Mech. Anal.* 164(2002), pp.213-259.

[8] Zhikov V. V., Passage to the limit in nonlinear variational problems, *Mat. Sb.* 183 (1993), pp.47-84 (in Russian), translation in *Russian Acad. Sci. Sb. Math.* 76(2)(1993), pp.427-359.

[9] Antontsev S., Shmarev S., Anisotropic parabolic equations with variable nonlinearity, *Publ. Mat.* 53(2009), pp.355-399.

[10] Zhikov V. V., On the density of smooth functions in Sobolev-Orlicz spaces, *Otdel. Mat. Inst. Steklov. (POMI)* 310(2004), pp.67-81, translation in J. Math. Sci. (N.Y.) 132(2006), pp.285-294.

[11] Fan X. L., Zhao D., On the spaces $L^{p(x)}(\Omega)$ and $W^{m,p(x)}$, *J. Math. Anal. Appl.* 263(2001), pp.424-446.

[12] Kovcik O., Rökosnäk J., On spaces $L^{p(x)}$ and $W^{k,p(x)}$, *Czechoslovak Math. J.* 41(1991), pp.592-618.

[13] Peng L. Q., Non-uniqueness for the p-harmonic heat flow with potential into homogeneous spaces, *Chinese Ann. Math. Ser. A.* 27(2006), pp.442-448.(in Chinese), translation in *Chinese J. Contemp. Math.* 27(2006), pp.231-238.

[14] Zhang C., Zhuo S., Xue X., Global gradient estimates for the parabolic $p(x,t)$ -Laplacian equation, *Nonlinear Analysis.* 105(2014), pp. 86-101.

[15] Yao F., Holder regularity for the general parabolic $p(x,t)$-Laplacian equations, *Nonlinear Differential Equations and Applications.* DOI 10.1007/s00030-014-0277-y, 2014.

[16] Berger M., Nonlinearity and functional analysis, in: *Lectures on Nonlinear Problems in Mathematical Analysis, Pure and Applied Mathematics* Academic Press, New York-London, 1977.

[17] Ladyzhenskaja O. A., Solonikov V. A., Ural'ceva N. N., *Linear and Quasilinear Equations of Parabolic Type*, American Mathematical Society, Providence, Rhode Island, 1968.

[18] Si X., Zhan H., The weak solutions to doubly nonlinear diffusion equation with convection term, *WSEAS Transactions on Mathematics*, 13(2014), pp. 416-427.

[19] Zhan H., Solutions to a convection diffusion equation, *Chinese Journal of contemporary mathematics*, 34(2)(2013), pp. 179-200.

[20] Zhan H., Xu B., The asymptotic behavior of the solution of a doubly degenerate parabolic equation with the convection term, *Journal of inequalities and applications*, 120(1)(2012), pp. 1-16.

[21] Zhan H., The nonexistence of the solution for quasilinear parabolic equation related to the p-Laplacian, *WSEAS Transactions on Mathematics*, 11(8)(2012), pp. 695-704.

[22] Zhan H., Solution to nonlinear parabolic equations related to the p-Laplacian, *Chinese Annals of Mathematics*, Series B, 33(4)(2012), pp. 767-782.

[23] Xie Q., Zhan H.,The singular diffusion equation with boundary begeneracy, *WSEAS Transactions on Mathematics*, 11(2)(2012), pp. 125-134.

[24] Zhan H., The self-similar solutions of a diffusion equation, *WSEAS Transactions on Mathematics*, 11(4)(2012), pp. 345-355.

[25] Zhan H., The asymptotic behavior of a doubly nonlinear parabolic equation with a absorption term related to the gradient, *WSEAS Transactions on Mathematics*, 10(7)(2011), pp. 229-238.

[26] Zhan H., Harnack estimates for weak solutions of a singular parabolic equation, *Chinese Annals of Math., Ser. B*, 32(2011), pp. 397-416.

[27] Zhan H., The asymptotic behavior of solutions for a class of doubly degenerate nonlinear parabolic equations, *J. Math. Anal. Appl.*, 370(2010), pp. 1-10.

[28] Zhan H., Gao F., A Note on the regularity of solutions for non-Newtonian filtration equations, *Southeast Asian Bulletin of Math.*, 30(2006), pp. 989-993.

In: Evolutionary Processes and Applications
Editor: G. M. N'Guérékata
ISBN: 978-1-53616-350-6

Chapter 11

ON A CRITERION FOR REGULARITY OF LERAY WEAK SOLUTIONS TO THE NAVIER - STOKES EQUATIONS INVOLVING ONE DIRECTIONAL DERIVATIVE

Ngo Van Giang[1,*] ***and Nguyen Minh Tri***[2,†]
[1]Faculty of Basic Sciences, Thai Nguyen University of Technology, Thai Nguyen, Vietnam
[2]Institute of Mathematics, Vietnam Academy of Science and Technology, Hanoi, Vietnam

Abstract

In this chapter we obtain sufficient conditions for the regularity of Leray solutions of Navier-Stokes equations (NSE) in three dimensional space. We prove that if the third derivative of the velocity $\frac{\partial u}{\partial x_3}$ belongs to the space $L_t^{s_0} L_x^{r_0}$, where $\frac{2}{s_0} + \frac{3}{r_0} \leq \frac{5}{3} + \frac{10}{3(9r_0-2)}, \quad r_0 \geq 3$ then the solution is regular.

Keywords: Navier - Stokes equations, regularity criterion, energy inequality, weak solution, one velocity component

---

*Corresponding Author's Email: ngogiangtcn@gmail.com.
†Corresponding Author's Email: triminh@math.ac.vn.

1. INTRODUCTION

In this chapter, we study the Cauchy problem for incompressible Navier - Stokes equations(NSE) in $\mathbb{R}^3$

$$\frac{\partial u_i}{\partial t} + \sum_{j=1}^{3} u_j \frac{\partial u_i}{\partial x_j} - \Delta u_i + \frac{\partial p}{\partial x_i} = 0 \text{ in } (0,T) \times \mathbb{R}^3, i = \overline{1,3}, \quad (1.1)$$

$$\operatorname{div}(u) = \sum_{i=0}^{3} \frac{\partial u_i}{\partial x_i} = 0 \text{ in } (0,T) \times \mathbb{R}^3, \quad (1.2)$$

$$u(0,x) = u_0(x) \text{ in } \mathbb{R}^3, \quad (1.3)$$

where $u : (0,T) \times \mathbb{R}^3 \longrightarrow \mathbb{R}^3$ is the velocity field, $p : (0,T) \times \mathbb{R}^3 \longrightarrow \mathbb{R}$ is the pressure. We recall the definitions of the spaces $C^\infty_{0,\sigma}(\mathbb{R}^3)$, $W^{1,2}_{0,\sigma}(\mathbb{R}^3)$, $L^2_\sigma(\mathbb{R}^3)$

$$C^\infty_{0,\sigma}(\mathbb{R}^3) = \{u \in C^\infty_0(\mathbb{R}^3), \operatorname{div}(u) = 0\},$$

$$W^{1,2}_{0,\sigma}(\mathbb{R}^3) = \text{ the closure of } C^\infty_{0,\sigma}(\mathbb{R}^3) \text{ in the topology } H^1_0(\mathbb{R}^3),$$

$$L^2_\sigma(\mathbb{R}^3) = \text{ the closure of } C^\infty_{0,\sigma}(\mathbb{R}^3) \text{ in the topology } L^2(\mathbb{R}^3).$$

The space $W^{1,2}_{0,\sigma}(\mathbb{R}^3)$ is a Hilbert space with the scalar product

$$((u,v)) = \sum_{i=1}^{3} (D_i u, D_i v) + (u,v).$$

The norm in $W^{1,2}_{0,\sigma}(\mathbb{R}^3)$ is defined by

$$\|u\|_{W^{1,2}_{0,\sigma}(\mathbb{R}^3)} = \|u\|_{1,2} := \Big(\sum_{|\alpha|\le 1} |D^\alpha u(x)|_2^2 \, dx\Big)^{\frac{1}{2}}.$$

The space $W^{1,2}_{0,\sigma}(\mathbb{R}^3)$ is contained and dense in $L^2_\sigma(\mathbb{R}^3)$, and the injection is continuous. Let $W^{-1,2}_{0,\sigma}(\mathbb{R}^3)$ denote the dual space of $W^{1,2}_{0,\sigma}(\mathbb{R}^3)$. By the Riesz representation theorem we have

$$W^{1,2}_{0,\sigma}(\mathbb{R}^3) \subset L^2_\sigma(\mathbb{R}^3) \subset W^{-1,2}_{0,\sigma}(\mathbb{R}^3).$$

Definition 1. *A vector field*

$$u \in L^\infty(0,T; L^2_\sigma(\mathbb{R}^3)) \cap L^2_{loc}(0,T; W_0^{1,2}(\mathbb{R}^3))$$

is called a weak solution of the problem (1.1) - (1.3) if the relation

$$\sum_{k=1}^{3} \int\int u_k(\partial_t w_k + \Delta w_k) + \sum_{j,k=1}^{3} \int\int u_j u_k \partial_j w_k + \sum_{k=1}^{3} \int u_k(x,0) w_k(x,0) dx = 0$$

is satisfied for all divergence - free $w \in C_0^\infty(\mathbb{R}^3 \times [0,T))$.

It is well known that for $u_0 \in L^2_\sigma(\mathbb{R}^3)$ problem (1.1)-(1.3) possesses at least one global weak solution u satisfying the energy inequality

$$\frac{\|u(t)\|_2^2}{2} + \int_0^t \|\nabla u(\tau)\|_2^2 \leq \frac{\|u_0\|_2^2}{2}$$

for every $t \geq 0$. A weak solution u satisfying the energy inequality is called a Leray weak solution.

We say that a Leray weak solution is regular if $u \in C^\infty((0,T) \times \mathbb{R}^3)$. It is well-known that if the above Leray weak solution u satisfies

$$u \in L^\infty(0,T; H^1(\mathbb{R}^3)) \cap L^2(0,T; H^2(\mathbb{R}^3)),$$

then such solution is regular. A vector field

$$u \in L^\infty(0,T; L^2_\sigma(\mathbb{R}^3)) \cap L^2_{loc}(0,T; W_0^{1,2}(\mathbb{R}^3))$$

is called a weak solution of NSE if the relation

$$\sum_{k=1}^{3} \int\int u_k(\partial_t w_k + \Delta w_k) + \sum_{j,k=1}^{3} \int\int u_j u_k \partial_j w_k + \sum_{k=1}^{3} \int u_k(x,0) w_k(x,0) dx = 0$$

is satisfied for all divergence - free $w \in C_0^\infty(\mathbb{R}^3 \times [0,T))$.

There are many known sufficient conditions on the velocity which guarantee that the solution is regular for all time. The first such criterion is usually referred to as the Prodi-Serrin condition saying that if a weak solution u additionally belongs to $L^t(0,T; L^s(\mathbb{R}^3)), \frac{2}{t} + \frac{3}{s} = 1, s \in [3,\infty)$, then the solution is regular and is unique in the class of all weak solutions satisfying the energy inequality. A related well-known sufficient condition is due to Beirão da Veiga [1]. Namely, if $\nabla u \in L^t(0,T; L^s(\mathbb{R}^3))$, $\frac{2}{t} + \frac{3}{s} = 2, s \in [\frac{3}{2}, \infty)$, then u is regular.

Recently, there have been many results in which requirement on one component ∇u_3 or on one directional derivative $\frac{\partial u}{\partial x_3}$ guarantees the regularity of weak solutions. In [2] and [3] the condition $\frac{\partial u}{\partial x_3} \in L_t^{s_0} L_x^{r_0}$ on a local time interval with $\frac{2}{s_0} + \frac{3}{r_0} \leq \frac{3}{2}$ and $2 \leq r_0 \leq \infty$ implies regularity. Also, Kucavica and Ziane [4] proved that $\frac{\partial u}{\partial x_3} \in L_t^{s_0} L_x^{r_0}$, where $\frac{2}{s_0} + \frac{3}{r_0} \leq 2$ with $\frac{9}{4} \leq r_0 \leq 3$, implies regularity. In 2008, Cao [5] extended regularity condition

$$\frac{\partial u}{\partial x_3} \in L_t^p L_x^q, \quad \frac{2}{p} + \frac{3}{q} = 2, \quad \frac{27}{16} \leq q \leq 3.$$

Very recently, Zhang [6] shows that the condition

$$\frac{\partial u}{\partial x_3} \in L_t^p L_x^q, \quad \frac{2}{p} + \frac{3}{q} = 2, \quad \frac{3\sqrt{37}}{4} - 3 \leq q \leq 3$$

could ensure the regularity of the solution.

Here, we improve the results of [2] and [3] in case $r_0 \geq 3$. We will show that if $\frac{\partial u}{\partial x_3} \in L_t^{s_0} L_x^{r_0}$ where $\frac{2}{s_0} + \frac{3}{r_0} \leq \frac{5}{3} + \frac{10}{3(9r_0 - 2)}$, $r_0 \geq 3$ then a Leray weak solution u is regular.

2. THE REGULARITY FOR WEAK SOLUTIONS

Lemma 2. *Let u be a sufficiently smooth divergence-free vector function in $\mathbb{R}^3$ with sufficient decay at infinity. Then,*

$$\sum_{i,j=1}^{2} \int_{\mathbb{R}^3} u_i \partial_i u_j \Delta_2 u_j dx = \frac{1}{2} \sum_{i,j=1}^{2} \int_{\mathbb{R}^3} \partial_3 u_3 \partial_i u_j \partial_i u_j dx$$
$$- \int_{\mathbb{R}^3} \partial_3 u_3 \partial_1 u_1 \partial_2 u_2 dx + \int_{\mathbb{R}^3} \partial_3 u_3 \partial_2 u_1 \partial_1 u_2 dx,$$

where $\Delta_2 = \partial_{11} + \partial_{22} = \partial^2/\partial x_1^2 + \partial^2/\partial x_2^2$ is two dimensional Laplacian.

Proof. See, e.g., Ref. [4] □

Lemma 3. *(Troisi) [7]. Let $p_i \in (1, \infty), q \in (1, \infty), n \geq 2$. Let $\sum_{i=1}^{n} \frac{1}{p_i} = \frac{q+n}{q}$. Then there exists a constant $C > 0$ such that for every $v \in W^{1,p_i}$*

$$\|v\|_q \leq C \prod_{i=1}^{n} \|\partial_i v\|_{p_i}^{\frac{1}{n}}.$$

Proof. See, e.g., Ref. [7] □

Now we state and prove the mail result of our note

Theorem 4. *Let* $u_0 \in H^1(\mathbb{R}^3)$*. Assume that* u *is a Leray weak solution of the problem (1.1)-(1.3). Let* $T > 0$, $\partial_3 u \in L_t^{s_0} L_x^{r_0}(\mathbb{R}^3 \times I)$ *with* $I \subset (0, T)$, *where*

$$\frac{2}{s_0} + \frac{3}{r_0} \leq \frac{5}{3} + \frac{10}{3(9r_0 - 2)}, \quad 3 \leq r_0,$$

then the Leray weak solution u *is regular on* $(0, T)$.

Proof. Let $T_{max} = \text{Sup}\{\tau > 0; \text{u is regular on}(0, \tau)\}$. Fix an arbitrary $T' \in (0, T)$ such that $T' \leq T_{max}$. We claim that the solution u cannot become singular at T'; this clearly implies $T_{max} = T$. Find $t_1 \in (0, T')$ such that

$$\|\partial_3 u\|_{L_t^{s_0} L_x^{r_0}(\mathbb{R}^3 \times (t_1, T'))} \leq \epsilon,$$

where ϵ is a sufficient small constant to be specified below. Then let $t_2 \in (t_1, T')$ be fixed but arbitrary, and let $t \in (t_1, t_2)$.
Set

$$J = \sum_{i=1}^{2} \|\nabla u_i\|_{L_t^\infty L_x^2(\mathbb{R}^3 \times (t_1, t))} + \|\Delta u_i\|_{L_t^2 L_x^2(\mathbb{R}^3 \times (t_1, t))},$$

$$K = \|\nabla u_3\|_{L_t^\infty L_x^2(\mathbb{R}^3 \times (t_1, t))} + \|\Delta u_3\|_{L_t^2 L_x^2(\mathbb{R}^3 \times (t_1, t))}.$$

Componentwise, the NSE may be rewritten as

$$\frac{\partial u_k}{\partial t} - \Delta u_k + \sum_{j=1}^{3} u_j \partial_j u_k + \partial_k p = 0 \quad (NSE_k),$$

where $k = 1, 2, 3$.
Now, for $k = 1, 2$, we multiply the equation NSE_k by $-\Delta u_k$, integrate over $\mathbb{R}^3 \times (t_1, t)$, and add them together. We obtain

$$\sum_{k=1}^{2} \int\int \frac{\partial u_k}{\partial t}(-\Delta u_k) + \sum_{k=1}^{2} \Delta u_k \Delta u_k = \sum_{k=1}^{2}\sum_{j=1}^{3} \int\int u_j \partial_j u_k \Delta u_k + \sum_{k=1}^{2} \int\int \partial_k p \Delta u_k$$

$$= I_1 + I_2 + I_3 + I_4,$$

where

$$I_1 = \sum_{j,k=1}^{2} \int\int u_j \partial_j u_k \Delta_2 u_k, \quad I_2 = \sum_{j,k=1}^{2} \int\int u_j \partial_j u_k \partial_{33} u_k,$$

$$I_3 = \sum_{k=1}^{2} \int\int u_3 \partial_3 u_k \Delta u_k, \quad I_4 = \sum_{k=1}^{2} \int\int \partial_k p \Delta u_k.$$

In the rest of the proof we will use the letter C for a universal constant which may very its value but it is independent of t. In [4] (with slight modifications) it was proved that

$$|I_1|, |I_2| \le C\epsilon J^2.$$

By using the Hölder, Sobolev and Trosi inequalities, we will estimate the term I_3.

Since $3 \le r_0$, we get

$$\begin{aligned}
|I_3| &= \left| \sum_{k=1}^{2} \int\int u_3 \partial_3 u_k \Delta u_k \right| \le \sum_{k=1}^{2} \int \|\partial_3 u\|_{r_0} \|\Delta u_k\|_2 \|u_3\|_{\frac{2r_0}{r_0-2}} \\
&\le \sum_{k=1}^{2} \int \|\partial_3 u\|_{r_0} \|\Delta u_k\|_2 \|u_3\|_2^{\frac{3r_0-8}{3r_0-2}} \|u_3\|_{3r_0}^{\frac{6}{3r_0-2}} \\
&\le \sum_{k=1}^{2} \int \|\partial_3 u\|_{r_0} \|\Delta u_k\|_2 \|u_3\|_2^{\frac{3r_0-8}{3r_0-2}} \|\nabla_h u_3\|_2^{\frac{4}{3r_0-2}} \|\partial_3 u_3\|_{r_0}^{\frac{2}{3r_0-2}} \\
&\le \sum_{k=1}^{2} \|\partial_3 u\|_{L^{s_0}(L^{r_0})}^{\frac{3r_0}{3r_0-2}} \|\Delta u_k\|_{L^2(L^2)} \|u\|_{L^\infty(L^2)}^{\frac{3r_0-8}{3r_0-2}} \|\nabla u_3\|_{L^\infty(L^2)}^{\frac{4}{3r_0-2}-b} \|\nabla u\|_{L^2(L^2)}^{b} \\
&\le C\epsilon^{\frac{3r_0}{3r_0-2}} J K^{\frac{4}{3r_0-2}-b},
\end{aligned}$$

where $\nabla_h u_3 := (\nabla_1 u_3, \nabla_2 u_3)$ and $\frac{1}{2} + \frac{3r_0}{(3r_0-2)s_0} + \frac{b}{2} = 1$.

Using integration by parts, we rewrite the fourth term I_4 as

$$I_4 = -\sum_{k=1}^{2} \int\int \Delta p \partial_k u_k = -\int\int \Delta p(\partial_1 u_1 + \partial_2 u_2).$$

Here,

$$\Delta p = -\sum_{i,j=1}^{3} \partial_i u_j \partial_j u_i = -\sum_{i,j=1}^{2} \partial_i u_j \partial_j u_i - 2\sum_{i=1}^{2} \partial_3 u_i \partial_i u_3 - \partial_3 u_3 \partial_3 u_3.$$

Hence,

$$I_4 = -\sum_{i,j=1}^{2}\int\int \partial_i u_j \partial_j u_i \partial_3 u_3 - 2\sum_{i=1}^{2}\int\int \partial_3 u_i \partial_i u_3 \partial_3 u_3 - \int\int (\partial_3 u_3)^3$$

$$:= I_4^{(1)} + I_4^{(2)} + I_4^{(3)}.$$

In [4] the authors obtained

$$|I_4^{(1)}|, |I_4^{(3)}| \leq C\epsilon J^2.$$

Next we consider the term $I_4^{(2)}$

$$I_4^{(2)} = 2\sum_{i=1}^{2}\int\int \partial_{i3} u_i u_3 \partial_3 u_3 + 2\sum_{i=1}^{2}\int\int \partial_3 u_i u_3 \partial_{i3} u_3$$

$$= 2\sum_{i=1}^{2}\int\int \partial_{i3} u_i u_3 \partial_3 u_3 - 2\sum_{i,j=1}^{2} \partial_3 u_i u_3 \partial_{ij} u_j.$$

Using estimate similar to the term I_3, and the second derivative estimate

$$\|\partial_{ij} u_k\|_{L^2(L^2)} \leq C\|\Delta u_k\|_{L^2(L^2)}$$

we have

$$|I_4^{(2)}| \leq C\epsilon^{\frac{3r_0}{3r_0-2}} J K^{\frac{4}{3r_0-2}-b}$$

with $r_0 \geq 3, b \geq 0$.

Combining the estimates for I_1, I_2, I_3 and I_4 we get

$$J^2 \leq C + C\epsilon J^2 + C\epsilon^{\frac{3r_0}{3r_0-2}} J K^{\frac{4}{3r_0-2}-b},$$

Choosing ϵ small enough, we derive

$$J \leq C + C\epsilon K^{\frac{4}{3r_0-2}-b} \tag{2.1}$$

with $r_0 \geq 3, b \geq 0$.

Next, multiplying the equation NSE_3 by $-\Delta u_3$, integrating over $\mathbb{R}^3 \times (t_1, t)$, we get

$$\int\int \frac{\partial u_3}{\partial t}(-\Delta u_3) + \int\int \Delta u_3 \Delta u_3 = \sum_{j=1}^{3}\int\int u_j \partial_j u_3 \Delta u_3 + \int\int \partial_3 p \Delta u_3$$

$$= N_1 + N_2 + N_3,$$

where

$$N_1 = \sum_{j=1}^{2} \int \int u_j \partial_j u_3 \Delta u_3, \quad N_2 = \int \int u_3 \partial_3 u_3 \Delta u_3, \quad N_3 = \int \int \partial_3 p \Delta u_3$$

Note that

$$N_3 = -I_4.$$

Next we estimate N_1

$$\begin{aligned} |N_1| &= \Big| \sum_{j=1}^{2} \int \int u_j \partial_j u_3 \Delta u_3 \Big| \le \sum_{j=1}^{2} \int \|\Delta u_3\|_2 \| u_j\|_{3r_0} \|\partial_j u_3\|_{s_1} \\ &\le \sum_{j=1}^{2} \int \|\Delta u_3\|_2 \|\partial_1 u_j\|_2^{1/3} \|\partial_2 u_j\|_2^{1/3} \|\partial_3 u_j\|_{r_0}^{1/3} \|\partial_j u_3\|_2^{\frac{6-s_1}{2s_1}} \|\partial_j u_3\|_6^{\frac{3s_1-6}{2s_1}} \\ &\le \sum_{j=1}^{2} \|\Delta u_3\|_{L^2(L^2)}^{\frac{5s_1-6}{2s_1}} \|\nabla u_j\|_{L^\infty(L^2)}^{\frac{2}{3}} \|\partial_3 u\|_{L^{s_0}(L^{r_0})}^{\frac{1}{3}} \|\nabla u_3\|_{L^\infty(L^2)}^{\frac{6-s_1}{2s_1}-a} \|\nabla u\|_{L^2(L}^{a} \\ &\le C\epsilon^{\frac{1}{3}} K^{2-a} J^{\frac{2}{3}}, \end{aligned}$$

where $\frac{5s_1-6}{4s_1} + \frac{1}{3s_0} + \frac{a}{2} = 1$ and $\frac{1}{3r_0} + \frac{1}{s_1} = \frac{1}{2}$.
Finally, using Höllder's inequality we derive a bound for N_2

$$|N_2| = \Big| \int \int u_3 \partial_3 u_3 \Delta u_3 \Big| \le C\epsilon K^2.$$

Chose $a = \frac{8}{3(3r_0-2)} - \frac{2b}{3}$. Using the estimates for N_1, N_2, and N_3 we conclud that

$$K^2 \le C + C\epsilon J^2 + C\epsilon K^2 + C\epsilon^{\frac{1}{3}} K^{2-a} J^{\frac{2}{3}} + C\epsilon J K^{\frac{4}{3r_0-2}-b}.$$

Hence

$$K^2 \le C + C\epsilon K^{2(\frac{4}{3r_0-2}-b)} + C\epsilon K^2 + C\epsilon^{\frac{1}{3}} K^{2-\frac{2}{3}(\frac{4}{3r_0-2}-b)} K^{\frac{2}{3}(\frac{4}{3r_0-2}-b)} + C\epsilon K^{\frac{4}{3r_0-2}-b} K^{\frac{4}{3r_0-2}-} \quad (2.$$

Choosing ϵ small enough, from (2.2) we obtain

$$K \le C.$$

Together with (2.1) it implies that $J \leq C$. Thus,

$$u \in L^{\infty}(t_1, t_2; H^1(\mathbb{R}^3)) \cap L^2(t_1, t_2; H^2(\mathbb{R}^3)).$$

Therefore, the solution cannot become singular at T'. This completes the proof. □

Acknowledgments

This research is funded by Vietnam National Foundation for Science and Technology Development (NAFOSTED) under grant number 101.02-2017.21.

References

[1] Beirão da Veiga, H., A new regularity class for the Navier-Stokes equations in $\mathbb{R}^n$. *Chin. Ann. Math.*, Ser. B 16 (1995), no. 4, pp. 407-412.

[2] Neustupa, J., Novotný, A. and Pene, P., An interior regularity of a weak solution to the Navier-Stokes equations in dependence on one component of velocity. *Topics in mathematical fluid mechanics, Quad. Mat.*, 10, Dept. Math., Seconda Univ. Napoli, Caserta, 2002, pp. 163-183.

[3] Pokorný, M., On the result of He concerning the smoothness of solutions to the Navier-Stokes equations. *Electron. J. Differential Equations* 2003, no. 11 , pp. 1-8.

[4] Kucavica, I. and Ziane, M., Navier-Stokes equations with regularity in one direction. *J. Math. Phys.* 48 (2007), no. 6, 10 pp.

[5] Cao, C. and Titi, E. S., Regularity criteria for the three dimensional Navier-Stokes equations. *Indiana Univ. Math. J.* 57 (2008), pp. 2643-2661.

[6] Zhang, Z., An improved regularity criterion for the Navier-Stokes equations in terms of one directional derivative of the velocity field. *Bull. Math. Sci.* 8 (2018), no. 1, pp. 33-47.

[7] Troisi, M., Embedding theorem for non - isotropic Sobolev Spaces. *Ric. Mat.* 18 (1969), pp. 3-24.

ABOUT THE EDITOR

Professor Gaston Mandata N'Guérékata
University Distinguished Professor of Mathematics,
The World Academy of Sciences (TWAS) Research Professor,
School of Computer, Mathematical and Natural Sciences,
Morgan State University, Baltimore, MD, US
Email: Gaston.N'Guerekata@morgan.edu

Professor Gaston M. N'Guérékata is a University Distinguished Professor and Associate Dean at Morgan State University in Baltimore, Maryland, USA. He earned a B.S, M.S and Ph. D. degrees in Mathematics from the University of Montreal in Canada. His dissertation on abstract differential equations was completed under the supervision of Professor Samuel Zaidman in 1980. Afterwards, he attended the University of California, Berkeley, through a postdoctoral program. His research areas include functional analysis, abstract harmonic analysis, spectral theory of bounded functions, functional and integro-differential equations, the asymptotic behavior of solutions of evolution equations, and both fuzzy and fractional differential equations. He also served as

Dean, Vice-Rector, and Rector a.i of the University of Bangui in his home coun- try of Central African Republic, where he held several high government posi- tions. He joined Morgan State University in 1996 as a Lecturer, then Associat Professor with tenure in 1997, and intensified his research considerably.

Professor N'Guérékata has made important contributions in the field of al- most periodic and almost automorphic functions and their applications to evolu tion equations. His monographs on almost automorphy are cornerstones in thi area. Currently, we see an extensive growth of interest to this field thanks to Pro fessor N'Guérékata's works. He is the author of over 240 research publications including 10 monographs at the doctorate/research level published by Springer Worldscientific, Kluwer Academic, and Nova Science. Professor N'Guérékat is the Founder and Chief Editor of the International Journal of Evolution Equa tions and the Journal of Nonlinear Evolution Equations and Applications. He i an Associate Editor for Applicable Analysis and a member of editorial board of over 16 other international mathematical journals. In 2013, the Journal o Nonlinear Studies, edited by Cambridge Scientific in the United Kingdom, an Cubo, a Brazilian mathematics journal, simultaneously published two special is sues to celebrate his 60th birthday. He has 100+ collaborators across the worl and over 6300 citations and a h-index of 39 according to Google scholar.

Professor N'Guérékata is a Fulbright Scholar, a Fellow of the Worl Academy of Sciences (TWAS), and Fellow of the African Academy of Sci ences (AAS). For his contributions to research and higher education, he ha received numerous honors in the US, such as the Iva Jones Medallion, the D Santye Jean McIntire, II, International Award, and the NAFEO Noble Awar He also received the Legion of Honor from France, the Order of Merit, and th Academic Palms from the Central African Republic.

He was the Claytor-Wooddard Invited Lecturer at the Joint Mathematic Meetings in San Antonio, Texas in January 2006, and the Joaquin Bustoz J Lecturer at the Blackwell-Tapia Conference held in 2014 at the Institute of Pu and Applied Mathematics at University of California at Los Angeles. He wa also the keynote speaker at the International Conference on Dynamical Sy tems in Kolkata, India in 2012, the International Conference on Mathematic Analysis and Application in Modelling (ICMAAM-2018) in Kolkata, India, an received the Memento of Honor from the Indian Mathematical Society in 201 He gave numerous invited lectures at several International Conferences in th US, India, China, Morocco, Guadeloupe, Cameroon, Vietnam, Burkina Fas and Senegal. He has visited universities in many countries including Franc

Canada, Puerto Rico, India, China, Guadeloupe, Martinique, Senegal, Burkina Faso, and the US.

Recently, Professor N'Guérékata was appointed in many professional positions, including a TWAS Research Professor to advise doctoral candidates at the University of Bangui, a member of the TWAS Mathematical Sciences Advisory Committee and a member of the American Mathematical Society Committee on Publications.

INDEX

A

adaptation, 66
amplitude, 16
anomalous diffusion, 63
applied mathematics, 74, 152
asymptotics, 63, 68
atmosphere, vii, viii, 15, 16, 17, 30, 31, 32, 33, 34, 35
automobile, 61

B

Banach spaces, 172, 173, 174, 177
Belarus, 49, 59
boundary value problem, ix, 24, 27, 175, 176, 179
bounded linear operators, 38, 108
bounds, 62, 65
Burkina Faso, 77

C

calculus, 38, 46
Cauchy problem, viii, 37, 38, 42, 68, 108, 171, 201, 206
classes, ix, 105, 106, 107, 109, 124, 151, 172
closure, 50, 82, 155, 177, 206
combined effect, 16, 17, 19, 21
computing, 95, 113
conduction, vii, viii, 15, 16, 17, 21, 27, 30, 31, 32, 33, 181
conductivity, 16, 18, 19, 28, 33
conservation, 18, 62, 63, 64, 65, 73
contradiction, 58
convergence, 12, 38, 43, 91, 144, 145, 148, 163
cooling, 32, 33

D

damping, 33
decay, 21, 29, 30, 62, 63, 73, 208
decomposition, 69
degenerate, 202, 203
derivatives, 42, 52, 53, 65, 68, 70, 71
dichotomy, ix, 105, 106, 107, 110, 112, 118, 121, 122, 125
differential equations, 19, 20, 50, 52, 54, 56, 57, 59, 74, 106, 109, 111, 124, 152, 153, 170, 172, 173, 174
diffusion, 1, 12, 73, 74, 75, 102, 106, 121, 202
diffusion process, 106
diffusivities, 16, 30
diffusivity, 16, 17, 20, 21, 30
divergence, 74, 207, 208
dynamic viscosity, 18
dynamical systems, 59

E

electromagnetic, 176
energy, 19, 21, 30, 205, 207
entropy, 19, 23
equilibrium, viii, 2, 4, 17, 49, 51

evolution, vii, ix, 1, 32, 46, 74, 103, 105, 106, 107, 108, 110, 111, 112, 121, 122, 124, 125, 151, 152, 153, 154, 160, 168, 169, 171, 172, 173, 201

F

families, 95, 108
filtration, 203
fluid, 19, 20, 34, 200, 213
formula, 16, 22, 25, 27, 83, 121
functional analysis, 202

G

genetics, 12, 13
gravity, 33, 34
growth, 2, 13, 19, 94

H

height, 16, 17
Hilbert space, 38, 78, 82, 206
hyperbolicity, 122

I

identification, 102
identity, 38, 39, 84, 134, 137, 146
inequality, ix, 40, 44, 45, 55, 56, 57, 58, 65, 66, 67, 69, 70, 73, 79, 80, 81, 82, 85, 91, 107, 109, 113, 115, 116, 117, 119, 127, 129, 130, 131, 138, 139, 140, 142, 147, 156, 161, 162, 163, 164, 165, 166, 167, 177, 178, 182, 187, 188, 189, 190, 191, 195, 196, 197, 198, 199, 205, 207, 212
integration, 26, 64, 133, 145, 146, 173, 210
interference, 95, 96
invasions, 1, 13

L

laws, 63, 73
linear function, 106
linear systems, 149
Lyapunov function, 59

M

magnetic field, vii, viii, 15, 16, 17, 18, 19, 21, 27, 29, 30, 31, 34, 35
magnetic field effect, 34
magnitude, 27
manifolds, ix, 105, 106, 107, 109, 112, 123, 124, 125
mapping, 106, 107, 115, 120, 121, 125, 153
mass, 18, 64, 65
master equation, 37, 46
migration, vii, 1, 2
modelling, 93
models, vii, 1, 2, 13
modifications, 210
modulus, 155

N

Nash equilibrium, 130
null, viii, ix, 77, 79, 80, 96, 103, 127, 129, 130, 131, 141, 148, 149

O

optimization, 129, 149
oscillation, 19

P

partial differential equations, 74, 102, 106, 108, 129
PDEs, 74, 77
periodicity, 157, 174
physics, 34, 75
pollution, viii, 77, 93, 95, 96, 101, 102
population, vii, 1, 2, 12, 13, 106
probability, vii, 1, 2
propagation, vii, 12, 15
proposition, 39, 40, 43, 45, 85, 86, 87, 89, 90, 100, 189

Q

quantum optics, 46

R

radiation, 20
radius, 110, 112, 114
reading, 200
real numbers, 50
recall, 63, 65, 107, 108, 110, 153, 155, 206
requirement, 208
requirements, viii, 49
residues, 27
resolution, 131
restrictions, 109, 179
root, 56
roots, 55
rules, 176

S

SAP, 156, 158, 159, 160, 162
second generation, 2
seed/seedlings, 2, 13
semigroup, viii, 37, 38, 39, 40, 41, 42, 43, 44, 45, 65, 121
sensitivity, viii, 77, 79, 96, 97, 103
smoothing, 73
smoothness, 213
Sobolev inequality, 177
sound speed, 18
Southeast Asia, 172, 203
specific heat, 18
stability, viii, 34, 49, 50, 52, 54, 56, 57, 58, 59, 125
state, 38, 58, 80, 90, 95, 98, 100, 102, 129, 135, 171, 209
stroke, 129, 149
structure, 46, 47, 114, 132

T

techniques, ix, 151, 152, 179
temperature, 17, 18
topology, 206
transformations, 53, 114
translation, 156, 201
transmission, vii, 15, 16
traveling waves, vii, 1, 2, 3, 13

V

variables, 18, 52, 54, 85
vector, viii, 38, 61, 62, 64, 78, 97, 99, 125, 129, 156, 207, 208
velocity, x, 18, 23, 205, 206, 207, 213
viscosity, vii, viii, 15, 16, 17, 19, 20, 21, 27, 29, 30, 31, 34

Y

yield, 69, 72

Related Nova Publications

Quaternion Matrix Computations

Authors: Musheng Wei, Ying Li, Fengxia Zhang and Jianli Zhao

Series: Mathematics Research Developments

Book Description: In this monograph, the authors describe state-of-the-art real structure-preserving algorithms for quaternion matrix computations, especially the LU, the Cholesky, the QR and the singular value decomposition of quaternion matrices, direct and iterative methods for solving quaternion linear systems, generalized least squares problems, and quaternion right eigenvalue problems.

Hardcover ISBN: 978-1-53614-121-4
Retail Price: $160

Mathematical Modeling for the Solution of Equations and Systems of Equations with Applications. Volume I

Authors: Ioannis K. Argyros, Santhosh George, and Narayan Thapa

Series: Mathematics Research Developments

Book Description: This book is dedicated to the approximation of solutions of nonlinear equations using iterative methods. The study about convergence matter of iterative methods is usually based on two categories: semi-local and local convergence analysis.

Hardcover ISBN: 978-1-53613-361-5
Retail Price: $230